Combinatorial Nullstellensatz

Combinatorial Nullstellensatz

With Applications to Graph Colouring

Xuding Zhu
R. Balakrishnan

CRC Press

Taylor & Francis Group
Boca Raton London New York

CRC Press is an imprint of the
Taylor & Francis Group, an **informa** business

A CHAPMAN & HALL BOOK

First edition published 2022
by CRC Press
6000 Broken Sound Parkway NW, Suite 300, Boca Raton, FL 33487-2742

and by CRC Press
2 Park Square, Milton Park, Abingdon, Oxon, OX14 4RN

Library of Congress Cataloging-in-Publication Data

ISBN: [978-0-367-68694-9] (hbk)
ISBN: [978-1-003-13867-9] (ebk)
ISBN: [978-0-367-68701-4] (pbk)

Typeset in CMR10
by KnowledgeWorks Global Ltd.

The first author (XZ) dedicates this book to his wife Tong Zhang.

The second author (RB) dedicates this book to Professor N. Ramanujam, former Head of the Department of Mathematics, Bharathidasan University.

Contents

Contents

Preface

Combinatorial Nullstellensatz (abbreviated as CNS), first introduced by Noga Alon in 1990, is a celebrated theorem in Algebra
which has been used to tackle combinatorial problems in diverse
areas of mathematics. Suppose $f(\mathbf{x}) \in \mathbb{F}[\mathbf{x}]$ is a polynomial with
coefficients in a field $\mathbb{F}$. CNS asserts that if the coefficient of a highest degree monomial $\prod_{i=1}^{n} x_i^{t_i}$ in the expansion of $f(\mathbf{x})$ is nonzero,
then any grid $S_1 \times S_2 \times \ldots \times S_n$ with $S_i \subseteq \mathbb{F}$ and $|S_i| = t_i + 1$,
contains a nonzero point of $f(\mathbf{x})$. To apply CNS to solve a com-
binatorial problem, we construct an appropriate polynomial $f(\mathbf{x})$
so that the solutions of the combinatorial problem correspond to
nonzero points of the polynomial. Then we show that the coefficient of a certain monomial of $f(\mathbf{x})$ is non-zero and invoke CNS to
conclude the existence of a solution to the combinatorial problem.

Chapter 1 introduces some basic notations. Chapter 2 presents
some examples of applications of CNS to show the varieties of problems where CNS is applied. The main part of this book is devoted
to the application of CNS to graph colouring and labeling problems. For these applications, the step of constructing a polynomial
for the combinatorial problems is easy. We concentrate on methods used to show that the coefficient of a certain monomial in the
expansion of a polynomial is nonzero. In general, it is a difficult
problem to determine if the coefficient of a monomial of a polymo-
nial is nonzero or not. Nevertheless, some methods are developed
for this purpose for some classes of special polynomials. One of the
methods is the so called Alon–Tarsi orientation. An orientation D
of a graph G is an Alon–Tarsi orientation if the number of even
Eulerian subdigraphs of D differs from the number of odd Eulerian subdigraphs of D. For the polynomial $f_G(\mathbf{x})$ corresponding
to proper colouring of a graph G, Alon–Tarsi's Theorem asserts
that the coefficient of a given monomial in the expansion of f_G is
nonzero if and only if a corresponding orientation D of G is Alon–
Tarsi. So the task reduces to finding an appropriate orientation of
a given graph. In general, it is difficult to determine if an orienta-

tion is Alon–Tarsi. Nevertheless, this method is successfully used
to solve some interesting list colouring problems. In Chapter 3, we
present a family of such results. For example, if a graph D is bipartite, then any orientation D is an Alon–Tarsi orientation, as D has
no odd Eulerian subdigraph. A consequence of this observation is
that bipartite planar graphs are 3-choosable. Two other easy obser-
vations are that if the total number of Eulerian subdigraphs of D
is odd, then the number of even Eulerian subdigraphs is certainly
different from the number of odd Eulerian subdigraphs. So D is
an Alon–Tarsi orientation. If the total number of Eulerian subdigraphs of D is congruent to 2 modulo 4, and the number of even
Eulerian subdigraphs of D is even, then again D is an Alon–Tarsi
orientation. The former is used in showing that any grid graph is
3-choosable, and the latter is used to solve an open problem of
Erdős that every 4-regular graph that decomposes into a Hamiltonian cycle and a family of triangles is 3-colourable. Induction
is also frequently used in proving the existence of an appropriate
Alon–Tarsi orientation. Chapter 3 presents a few such proofs that
solve list colouring problems for planar graphs. In particular, this
method is used to show that every planar graph G has a matching
M such that $G - M$ is 4-choosable, and currently there is no other
proof of this result.

The second method for showing the coefficient of a certain
monomial is nonzero is to calculate the coefficient through an interpolation formula. Uwe Schauz and Michał Lasoń independently
proved an interpolation formula that expresses the coefficient of a
monomial $\prod_{i=1}^{n} x_i^{t_i}$ of $f(\mathbf{x})$ as a linear combination of the evalua-
tions of $f(\mathbf{x})$ on an arbitrary grid $S_1 \times S_2 \times \ldots \times S_n$ with $|S_i| = t_i+1$.
Such an interpolation formula is a generalization of CNS, because
if $f(\mathbf{x})$ has no nonzero point in a grid $S_1 \times S_2 \times \ldots \times S_n$ with
$|S_i| = t_i + 1$, then certainly the coefficient of $\prod_{i=1}^{n} x_i^{t_i}$ is zero. The
interpolation formula turns out to be quite useful in solving some
graph colouring problems. In Chapter 4, we present a result of
Schauz which asserts that if p is a prime, then the line graph of
K_{p+1} is p-choosable. Also it is shown that every r-edge colourable
planar graph is r-edge choosable and the line graphs of cycle powers are chromatic choosable. The former result was first proved by
Ellingham and Goddyn and the latter result was first proved by
Woodall, and we present new proofs of these results by using the
interpolation formula. Some other generalizations of CNS are also
discussed in this chapter.

The third method for calculating the coefficient of a certain monomial is by calculating the permanents of certain matrices. This method is particularly applied to the edge-weighting and total-weighting of graphs. The celebrated 1-2-3 conjecture asserts that if G is a connected graph on at least 3 vertices, then there is a labeling $f : E(G) \to \{1,2,3\}$ so that for any two adjacent vertices u and v, $\sum_{e \in E(u)} f(e) \neq \sum_{e \in E(v)} f(e)$. The analogous 1-2 conjecture asserts that for any graph G, there is a labeling $f : V(G) \cup E(G) \to \{1,2\}$ so that for any two adjacent vertices u and v, $\sum_{e \in E(u)} f(e) + f(u) \neq \sum_{e \in E(v)} f(e) + f(v)$. The list version of the 1-2-3 conjecture asserts that if L is a list assignment with $|L(e)| = 3$ for $e \in E(G)$, then there is a required labeling f with $f(e) \in L(e)$. Analogously, we have the list version of the 1-2 conjecture. CNS is used in most of the studies of the list version of these two conjectures. The coefficients of the corresponding polynomial turn out to be the permanents of matrices formed by columns of A_G and B_G, where A_G, B_G are matrices associated to a graph G. In Chapter 5, we present some results concerning the list version of the 1-2-3 conjecture and the 1-2 conjecture obtained by using this method. In particular, this method is used to show that complete graphs and complete bipartite graphs are $(1,3)$-choosable, trees with an even number of edges are $(1,2)$-choosable, and every graph is $(2,3)$-choosable.

CNS is now a widely used tool in the study of many combinatorial problems in a variety of areas in mathematics. This little book originates from the lecture notes of the first author given at a graduate student course and lecture notes the second author given at a workshop, with inputs from the first author. It is not intended to be a comprehensive survey of the application of this tool. Instead, it is intended to impress upon its readers of the usefulness of CNS and inspire them to further study and use this tool in their research.

Authors

Xuding Zhu is currently a Professor of Mathematics and also the Director of the Center for Discrete Mathematics at Zhejiang Normal University, China. His fields of interest are: Combinatorics and Graph Colouring. He published more than 260 research papers. He has served/is serving on the editorial board of *SIAM Journal on Discrete Mathematics, Journal of Graph Theory, European Journal of Combinatorics, Electronic Journal of Combinatorics, Discrete Mathematics, Contribution to Discrete Mathematics, Discussiones Mathematicae. Graph Theory, Bulletin of Academia Sinica, Indian Journal of Discrete Mathematics* and *Taiwanese Journal of Mathematics.*

R. Balakrishnan is currently an Adjunct Professor of Mathematics at Bharathidasan University, Triuchirappalli, India. His fields of interests are: Algebraic Combinatorics and Graph Colouring. He is an author of three other books, one in Graph Theory and the other two in Discrete Mathematics. He is also one of the founders of the Ramanujan Mathematical Society. In addition, he is currently an Editor-in-Chief of the *Indian Journal of Discrete Mathematics.*

Acknowledgements

The first author (XZ) thanks Daniel Cranston, Uwe Schauz and Tsai-Lien Wong for their careful reading and valuable comments on the first draft of the manuscript.

The second author (RB) thanks the authorities of Bharathidasan University, Tiruchirappalli for their support and his colleagues in the mathematics department for their encouragement which proved vital to this venture.

Chapter 1

Some Definitions and Notations

In this chapter, we present some basic definitions and notations. Some other definitions are given at the appropriate places of the book.

Definition 1.0.1 *A graph G consists of a vertex set $V(G)$ and edge set $E(G)$. A graph G is a finite graph if both $V(G)$ and $E(G)$ are finite sets; otherwise, G is an infinite graph.*

The edges of a graph G are unordered pairs $e = \{u, v\}$ of vertices in V. If $e = \{u, v\}$ is an edge of G, then e is incident at u and v. Sometimes we write $e = uv$ instead of $e = \{u, v\}$. When the graph G is clear from the context, we simply write V and E in place of $V(G)$ and $E(G)$, respectively. An edge of the form $\{v, v\}$ is called a loop at vertex v. Two edges of G having the same end vertices are called parallel edges or multiple edges of G. A graph without loops and without multiple edges is called a simple graph. A graph H is a subgraph of G if $V(H) \subseteq V(G)$ and $E(H) \subseteq E(G)$. A subgraph H of G is called a spanning subgraph of G if $V(H) = V(G)$.

Definition 1.0.2 *The number of edges of G incident at a vertex u of G is called the* degree *of u in G, and is denoted by $d_G(u)$ or simply by $d(u)$ when the underlying graph G is clear from the context. A loop at u contributes 2 to the degree of u.*

Definition 1.0.3 *A walk W in a graph G is an alternating sequence of vertices and edges of the form $v_1 e_1 v_2 e_2 \ldots v_i e_i v_{i+1} \ldots v_{p-1} e_{p-1} v_p$, where e_i is the edge $v_i v_{i+1}, 1 \leq i \leq p - 1$. A trail in G is a walk in G in which no edge is repeated. A path is a trail in which no vertex is repeated.*

Definition 1.0.4 *A walk (resp. trail, path) is closed if its initial and terminal vertices coincide. A closed path is called a cycle.*

Definition 1.0.5 *A graph G is* connected *if there is a path between any two distinct vertices of G; otherwise, G is disconnected.*

A component *of a graph G is a maximal connected subgraph of G. So if G is connected, it has just one component.*

Definition 1.0.6 *Given a graph G, its* line graph $L(G)$ *is defined as follows: Each vertex of $L(G)$ corresponds to an edge of G, and two vertices of $L(G)$ are adjacent in $L(G)$ if, and only if, the corresponding edges have a common endvertex in G. Therefore $|V(L(G))| = |E(G)|$. Moreover, $|E(L(G))| = \sum_{v \in V(G)} \binom{d(v)}{2} = \frac{1}{2} \sum_{v \in V(G)} (d_G(v))^2 - |E(G)|$.*

Definition 1.0.7 *A* digraph D *consists of a vertex set $V(D)$ and an arc set $A(D)$. The arcs of D, that is, the elements of $A(D)$, are ordered pairs (u, v) of vertices of D.*

The definitions of a directed path, a directed cycle etc., in a digraph D are all analogous to the definitions in the undirected case. We presume that the reader is familiar with the basics of groups, rings, integral domains, fields (both infinite and finite) and vector spaces. For more details about graphs, digraphs and fields, the reader may consult the books [7], [63] and [8].

Definition 1.0.8 *A graph G is* planar *if there exists a drawing of G in the plane in which no two edges intersect at a point other than a vertex of G, where each edge is a Jordan arc (that is, a simple arc in the plane). Such a drawing of a planar graph is called a* plane representation *of G. In this case, we also say that G has been embedded in the plane. A* plane graph *is a planar graph that has already been embedded in the plane.*

Definition 1.0.9 *Let P be the set of points of the plane minus the points of the plane graph G (that is, points belonging to the edges of G). Define $p \sim p'$ in P if and only if there exists a Jordan arc from p to p' in P. This relation is an equivalence relation Π on P. The equivalence classes of Π are called the* faces *of the plane graph G.*

Let G be a connected plane graph with n vertices, m edges and f faces. There is a nice classical relation connecting n, m and f due to Euler.

Theorem 1.0.10 *(Euler's Formula)　　For a connected plane graph G with n vertices, m edges and f faces.*

$$n - m + f = 2.$$

Let G be a graph and $S = \{1, 2, \ldots, k\}$. A *k-colouring* of G is a map c from $V(G)$ to S; it is *proper*, if adjacent vertices of G are mapped to distinct elements of S by c, that is, $uv \in E(G) \Rightarrow c(u) \neq c(v)$. The elements of S are called *colours*. The set of vertices which receive the same colour q is an independent subset V_q of $V(G)$ and is called a *colour class*.

Definition 1.0.11 *The* chromatic number *of a graph G, denoted by $\chi(G)$, is the least positive integer k such that G has a proper vertex colouring using k colours.*

Equivalently, the chromatic number $\chi(G)$ of G is the minimum size of a partition of the vertex set $V(G)$ of G into independent subsets of G.

Note that in considering proper vertex colourings of graphs, it is enough to restrict to simple graphs.

Definition 1.0.12 *A graph G is* bipartite *if its vertex set $V(G)$ can be partitioned into two independent subsets.*

Definition 1.0.13 *A list assignment of a graph G is a mapping L which assigns to each vertex v a set of permissible colours. An L-colouring of G is a colouring c of G such that $c(v) \in L(v)$ for each vertex v. We say that G is L-colourable if there exists a proper L-colouring of G.*

Definition 1.0.14 *A k-list assignment of G is a list assignment L of G with $|L(v)| = k$ for every vertex v of G. A graph G of is k-choosable if G is L-colourable for each k-list assignment L. The* choice number $ch(G)$ *of G is the minimum k for which G is k-choosable.*

If $ch(G) = k$, then by considering the list assignment L with $L(v) = \{1, 2, \ldots, k\}$ for every vertex v, we conclude that G is k-colourable. So $ch(G) \geq \chi(G)$. However, the difference $ch(G) - \chi(G)$ can be arbitrarily large as the following theorem shows:

Theorem 1.0.15 ((Erdős–Rubin–Taylor)[23]) *If $m = \binom{2k-1}{k}$, then the complete bipartite graph $K_{m,m}$ is not k-choosable.*

Proof. Let (X, Y) be the bipartition of the bipartite graph $G = K_{m,m}$. Let L be the list assignment that assigns the distinct k-subsets of the set $\{1, 2, 3, \ldots, 2k-1\}$ to the m vertices of X and also to the m vertices of Y. We shall show that G is not L-colourable.

Assume to the contrary that f is a proper L-colouring of G. Then f uses at least k colours for vertices in X; for otherwise, there is a k-set $S \subset \{1, 2, \ldots, 2k-1\}$ not used for vertices in X. But then, there is a vertex $v \in X$ with $L(v) = S$, a contradiction. Similarly, f uses at least k colours for vertices in Y. But there are $2k-1$ colours in total. So one colour is used for a vertex in X and a vertex in Y, and f is not a proper colouring. $\blacksquare$

Notation 1 $\mathbb{N} = \{1, 2, \ldots\}$ *stands for the set of natural numbers. Furthermore,* $\mathbb{N}_0 = \mathbb{N} \cup \{0\}$ *and* $\mathbb{N}_0^n$ *is the set of vectors* $(t_1, t_2, \ldots, t_n)$ *of length* n, *where* $t_i \in \mathbb{N}_0$ *for each* $i \in \{1, 2, \ldots, n\}$.

Notation 2 *A* grid *of dimension* n *is the Cartesian product of* n *non-empty finite sets, written as:* $\mathbf{S} = S_1 \times S_2 \times \cdots \times S_n$, *where* $\mathbf{S}$ *is given in bold. The* size $\|\mathbf{S}\|$ *of* $\mathbf{S}$ *is defined to be the integer vector* $(|S_1| - 1, |S_2| - 1, \ldots, |S_n| - 1)$.

Definition 1.0.16 *Suppose that* F *is a field and* $f(\mathbf{x}) = f(x_1, x_2, \ldots, x_n) \in F[x_1, x_2, \ldots, x_n]$ *is a polynomial. If* $f(\mathbf{x}) = \prod_{i=1}^{n} x_i^{t_i}$, *then* $f(\mathbf{x})$ *is called a* monomial. *We write* $\prod_{i=1}^{n} x_i^{t_i} = \mathbf{x}^{\mathbf{t}}$, *where* $\mathbf{t} = (t_1, t_2, \ldots, t_n) \in \mathbb{N}_0^n$. *By an* expansion *of a polynomial* $f(\mathbf{x})$, *we mean the expression of* $f(\mathbf{x})$ *as a linear combination of monomials. For* $\mathbf{t} \in \mathbb{N}^n$, *we denote by* $c_{f,\mathbf{t}}$ *the coefficient of* $\mathbf{x}^{\mathbf{t}}$ *in the expansion of* $f(\mathbf{x})$ *and let* $|\mathbf{t}| = \sum_{i=1}^{n} t_i$.

Definition 1.0.17 *Let* $f(\mathbf{x}) = f(x_1, x_2, \ldots, x_n) \in F[x_1, x_2, \ldots, x_n]$ *be a polynomial, and* $\mathbf{a} = (a_1, a_2, \ldots, a_n) \in F^n$. *If* $f(\mathbf{a}) = 0$, *then* $\mathbf{a}$ *is called a* zero point *of* f. *If* $f(\mathbf{a}) \neq 0$, *then* $\mathbf{a}$ *is called a* non-zero point *of* f.

Notation 3 $\mathbb{R}^n$ *stands for the* n-dimensional Euclidean space *(over* $\mathbb{R}$*). A* canonical basis *(or* standard basis*) for* $\mathbb{R}^n$ *over* $\mathbb{R}$ *is the set of* n *vectors* $\{e_1, e_2, \ldots, e_n\}$, *where* $e_1 = (1, 0, \ldots, 0), e_2 = (0, 1, \ldots, 0), \ldots, e_n = (0, 0, \ldots, 0, 1)$.

For sets A *and* B, *the set of points of* A *not in* B *is denoted by* $A \setminus B$. *The* disjoint union *of* A *and* B *is denoted by* $A \dot{\cup} B$. *For a graph* G *and a subset* M *of* $E(G)$, *we denote by* $G - M$ *the subgraph obtained by removing edges in* M *from* G.

We use boldface lower case letters to denote vectors whose coordinates are denoted by the same (but not boldface) lower case letter; for example, $\mathbf{a} = (a_1, a_2, \ldots, a_n)$ and $\mathbf{x} = (x_1, x_2, \ldots, x_n)$, where the dimension n is usually clear from the context.

Definition 1.0.18 *Let D be a digraph. For a vertex x, denote by $A_D(x)$ (respectively, $A_D^+(x), A_D^-(x)$) the set of arcs incident at x (respectively, the set of out-going arcs or incoming arcs at x), $d_D^+(x) = |A_D^+(x)|$ (respectively, $d_D^-(x) = |A_D^-(x)|$) is the out-degree (respectively, the in-degree) of x. Let $\Delta_D^+ = \max\{d_D^+(x) : x \in V(G)\}$ (respectively, $\Delta_D^- = \max\{d_D^-(x) : x \in V(G)\}$). We say that D is* Eulerian *if $d_D^+(x) = d_D^-(x)$ for each vertex x of D.*

Definition 1.0.19 *Let D be a digraph. An* Eulerian subdigraph *of D is an Eulerian digraph H with $V(H) = V(D)$ and $A(H) \subseteq A(D)$. An Eulerian subdigraph H of D is* even *(respectively,* odd*) if the number of arcs of H is even (respectively, odd).*

Definition 1.0.20 *We denote by $\mathcal{E}(D)$ the set of Eulerian subdigraphs of D. Let*

$$\mathcal{E}_e(D) := \{H \in \mathcal{E}(D) : H \text{ is even}\},$$
$$\mathcal{E}_o(D) := \{H \in \mathcal{E}(D) : H \text{ is odd}\}.$$

The difference *of D is defined as*

$$\text{diff}(D) = |\mathcal{E}_e(D)| - |\mathcal{E}_o(D)|.$$

Note that $\text{diff}(D)$ *can be negative. A digraph D is called* Alon–Tarsi *(shortened as AT) if* $\text{diff}(D) \neq 0$, *i.e.,* $|\mathcal{E}_e(D)| \neq |\mathcal{E}_o(D)|$.

Note that an Eulerian subdigraph H of D has the same vertex set as D; i.e., H is a spanning subdigraph of D. Also an Eulerian subdigraph H of D need not be connected (in contrast to the usual definition of Eulerian graphs). In particular, the empty subdigraph, which is the subdigraph having no arcs, of D is an even Eulerian subdigraph of D. So each digraph D has at least one even Eulerian subdigraph.

For arcs $a_1, \ldots, a_p, b_1, \ldots, b_q$ of D, we denote by

$$\mathcal{E}(D, a_1, \ldots, a_p, \overline{b}_1, \ldots, \overline{b}_q)$$

the set of Eulerian subdigraphs $H \in \mathcal{E}(D)$ whose arc set contains $\{a_1, \ldots, a_p\}$ and is disjoint from $\{b_1, \ldots, b_q\}$. For $\mathcal{E}' \subseteq \mathcal{E}(D)$, let

$$\mathcal{E}_e' = \{H \in \mathcal{E}' : |A(H)| \text{ is even}\},$$
$$\mathcal{E}_o' = \{H \in \mathcal{E}' : |A(H)| \text{ is odd}\},$$
$$\text{diff}(\mathcal{E}') = |\mathcal{E}_e'| - |\mathcal{E}_o'|.$$

Chapter 2

Combinatorial Nullstellensatz

2.1 Introduction

Combinatorial Nullstellensatz (abbreviated as CNS herein) is a landmark theorem in algebraic combinatorics, which is now a widely used tool in tackling combinatorial problems in diverse areas of mathematics. CNS was first introduced by Noga Alon in [4] wherein he has shown its applications in solving a variety of problems-providing new proofs to some of the classical results and solving new problems. The German word 'Nullstellensatz' may be roughly translated as 'zero location theorem' where the word zero refers to the zeros of a single polynomial or a set of polynomials. Alon's proof of CNS, as given in [4] is based on the classical theorem in Algebra: Hilbert's Nullstellensatz (HNS) (See for instance the book [43] for a proof), which is basic in Algebraic Geometry. However, our proof of CNS is the one given by Michałek [47] based on mathematical induction. We start by stating HNS.

Definition 2.1.1 *Let $\mathbb{F}$ be a field and let $f(\mathbf{x}) = f(x_1, x_2, \ldots, x_n)$ be a polynomial in $\mathbb{F}[x_1, x_2, \ldots, x_n]$. We say $\mathbf{x^t}$ is* a non-vanishing monomial *of f if $c_{f,\mathbf{t}} \neq 0$. The* degree of f *is*

$$\deg f = \max\{|\mathbf{t}| : c_{f,\mathbf{t}} \neq 0\}.$$

Theorem 2.1.2 (Hilbert's Nullstellensatz) *Let k be a field, and K its algebraic closure. Let I be an ideal in the polynomial ring $k[x_1, x_2, \ldots, x_n]$, and let $V(I)$ be the set of all points $\mathbf{x} = (x_1, x_2, \ldots, x_n) \in K^n$ such that $f(\mathbf{x}) = 0$ for every $f \in I$. If $p \in k[x_1, x_2, \ldots, x_n]$ is any polynomial that vanishes on all of $V(I)$ (i.e., $p(\mathbf{x}) = 0$ for each $\mathbf{x} \in V(I)$), then there exists a positive integer t such that $p^t \in I$, that is, p belongs to the radical of I.*

Recall that for subsets $S_1, S_2, \ldots, S_n$ of a field F, the set

$$\mathbf{S} := S_1 \times S_2 \times \cdots \times S_n = \{(a_1, a_2, \ldots, a_n) : a_i \in S_i\}$$

is an n-dimensional grid over F and the *size* $||\mathbf{S}||$ of $\mathbf{S}$ is the integer vector $(|S_1| - 1, |S_2| - 1, \ldots, |S_n| - 1)$.

Theorem 2.1.3 (Combinatorial Nullstellensatz (CNS [4]))
Let $\mathbb{F}$ be a field and let $f(\mathbf{x}) = f(x_1, x_2, \ldots, x_n)$ be a non-zero polynomial in $\mathbb{F}[x_1, x_2, \ldots, x_n]$. Suppose that $\mathbf{x^t}$ is a highest degree non-vanishing monomial of f. Then for any n-dimensional grid $\mathbf{S}$ of $\mathbb{F}$ with $||\mathbf{S}|| = \mathbf{t}$, there exists $\mathbf{s} \in \mathbf{S}$ for which $f(\mathbf{s}) \neq 0$. In other words, if $c_{f,||\mathbf{S}||} \neq 0$, then the grid $\mathbf{S}$ contains a non-zero point of f.

NOTE: Equivalently, we can as well replace $||\mathbf{S}|| = \mathbf{t}$ by $||\mathbf{S}|| \geq \mathbf{t}$, i.e., $|S_i| - 1 \geq t_i$ for $i = 1, 2, \ldots, n$.

Proof. (M. Michałek [47]) By induction on the degree of f.

Assume that $\deg f = 1$ so that

$$f(\mathbf{x}) = a_1 x_1 + a_2 x_2 + \cdots + a_n x_n + a_{n+1},$$

$a_i \in \mathbb{F}$ with at least one $a_i \neq 0$, say $a_1 \neq 0$. Assume that $S_1, S_2, \ldots, S_n \subseteq \mathbb{F}$ with $|S_1| = 2$, and $|S_i| = 1$ for $i = 2, 3, \ldots, n$. Assume that $S_1 = \{s_1, s_1'\}$ and $S_i = \{s_i\}$ for $i = 2, 3, \ldots, n$. Let $\mathbf{s} = (s_1, s_2, \ldots, s_n)$ and $\mathbf{s}' = (s_1', s_2, \ldots, s_n)$. Then $f(\mathbf{s}) - f(\mathbf{s}') = a_1(s_1 - s_1') \neq 0$. Hence at least one of $\mathbf{s}, \mathbf{s}'$ is a non-zero point of f.

Assume $f(\mathbf{x}) \in \mathbb{F}[x_1, x_2, x_3, \ldots, x_n]$, $\deg f = r \geq 2$ and the result holds for polynomials of degree $\leq r-1$. Suppose $\mathbf{x^t}$ is a highest degree non-vanishing monomial of f, where $\mathbf{t} = (t_1, t_2, \ldots, t_n)$ and $|\mathbf{t}| = r$. Thus $t_i \geq 0$ for each i and $t_1 + t_2 + \cdots + t_n = r$. Without loss of generality, assume that $t_1 > 0$. Suppose f does not satisfy the statement in the theorem, $S_1, S_2, \cdots, S_n$ are subsets of $\mathbb{F}$ with $|S_i| = t_i + 1$ and $f(\mathbf{s}) = 0$ for all $\mathbf{s} = (s_1, s_2, \ldots, s_n) \in S_1 \times S_2 \times \ldots \times S_n$.

Fix $a \in S_1$ and write

$$f(\mathbf{x}) = (x_1 - a)g(\mathbf{x}) + R(\mathbf{x}), \tag{2.1}$$

(that is, divide $f(\mathbf{x})$ by $(x_1 - a)$ getting $g(\mathbf{x})$ as quotient and $R(\mathbf{x})$ as remainder, where $R(\mathbf{x})$ does not contain x_1). Then $x_1^{t_1 - 1} x_2^{t_2} \ldots x_n^{t_n}$ is a non-vanishing monomial of $g(\mathbf{x})$ of highest degree.

For any $\mathbf{s} \in \{a\} \times S_2 \cdots \times S_n$, $f(\mathbf{s}) = 0$ by our assumption, which implies that $R(\mathbf{s}) = 0$. By virtue of (2.1), we get

$$R(\mathbf{s}) = 0, \quad \text{for every } \mathbf{s} \in S_2 \times S_3 \times \ldots \times S_n.$$

Thus both $f(\mathbf{x})$ and $R(\mathbf{x})$ vanish on the set

$$(S_1 - \{a\}) \times S_2 \cdots \times S_n,$$

and hence so does $g(\mathbf{x})$. But this contradicts our induction assumption on $g(\mathbf{x})$. ∎

Corollary 2.1.4 *Let $\mathbb{F}$ be a field and let $f(\mathbf{x}) = f(x_1, x_2, \ldots, x_n)$ be a non-zero polynomial in $\mathbb{F}[x_1, x_2, \ldots, x_n]$. Assume $\mathbf{S} = S_1 \times S_2 \times \ldots \times S_n$ and $\sum_{j=1}^{n}(|S_j| - 1) > \deg(f)$. Then $\mathbf{S}$ contains either no non-zero point or more than one non-zero point.*

Proof. Assume to the contrary that $\sum_{j=1}^{n}(|S_j| - 1) > \deg(f)$ and $\mathbf{S}$ contains exactly one non-zero point $\mathbf{s} = (s_1, s_2, \ldots, s_n)$ of f. Let

$$g(\mathbf{x}) = f(\mathbf{s}) \left(\prod_{i=1}^{n} \prod_{b \in S_i - \{s_i\}} (s_i - b) \right)^{-1} \prod_{i=1}^{n} \prod_{b \in S_i - \{s_i\}} (x - b).$$

Note that $g(\mathbf{x}) = 0$ for all $\mathbf{x} \in \mathbf{S} - \{\mathbf{s}\}$ and $g(\mathbf{s}) = f(\mathbf{s})$. Thus $(f - g)(\mathbf{x}) = 0$ for all $\mathbf{x} \in \mathbf{S}$, and $c_{f-g,||\mathbf{S}||} = c_{-g,||\mathbf{S}||} = -f(\mathbf{s}) \left(\prod_{i=1}^{n} \prod_{b \in S_i - \{s_i\}} (s_i - b) \right)^{-1} \neq 0$. This is a contradiction to Theorem 2.1.3. ∎

Corollary 2.1.5 *Assume that $\mathbb{F}$ is finite field with $|\mathbb{F}| = q$ and $P_1(\mathbf{x}), \ldots, P_m(\mathbf{x})$ are polynomials in $\mathbb{F}[x_1, x_2, \ldots, x_n]$ with $(q - 1) \sum_{j=1}^{m} \deg(P_j(\mathbf{x})) < \sum_{i=1}^{n}(|S_i| - 1)$. Then $\mathbf{S} = S_1 \times \ldots \times S_n$ contains either no common zero point or more than one common zero point of the P_j's.*

Proof. Let $P(\mathbf{x}) = \prod_{j=1}^{m}(1 - P_i(\mathbf{x})^{q-1})$. Then $P(\mathbf{x}) \neq 0$ if and only if $\mathbf{x}$ is a common zero point of all the P_j's. The assumption of this corollary implies that $\deg(P) < \sum_{i=1}^{n}(|S_i| - 1)$. So the conclusion follows from Corollary 2.1.4. ∎

CNS has applications to diverse areas of mathematics. In the book, we shall concentrate on application to graph colouring. However, in the remainder of this chapter, we present some applications of CNS to number theory, graph structure and geometry (hyperplanes in $\mathbb{R}^n$).

2.2 An application of CNS to additive number theory

We now present a proof due to Alon [4] of one of the celebrated classical theorems in additive number theory due to Cauchy and

Davenport as an application of CNS. This result was first proved by Cauchy and rediscovered by Davenport after a gap of 122 years!

Theorem 2.2.1 (Cauchy and Davenport) *Let p be a prime, and let A and B be two nonempty subsets of the field $\mathbb{Z}_p$. Let $C = \{x \in \mathbb{Z}_p : x = a + b \text{ for some } a \in A \text{ and } b \in B, a \neq b\}$. Then $|C| \geq \min\{p, |A| + |B| - 3\}$.*

Proof. The case $p = 2$ is trivial. So assume that $p \geq 3$.

Case 1 $p \leq |A| + |B| - 3$.

Take any $g \in \mathbb{Z}_p$. Then $|g - B| = |B|$ and $|A| + |g - B| \geq p + 3$. Hence $|A \cap (g - B)| \geq 3$. Let $a \in A \cap (g - B) - \{g/2\}$. (Note: $g/2$ exists in $\mathbb{Z}_p$ as $p(\geq 3)$ is an odd prime). Then $g = a + b$ for some $b \in B$ with $a \neq b$. So $g \in C$. As g is an arbitrary element of $\mathbb{Z}_p$), $C = \mathbb{Z}_p$, and the theorem is true in this case.

Case 2 $|A| + |B| - 3 < p$.

Assume the contrary to the statement of the theorem so that $|C| < \min(p, |A| + |B| - 3) = |A| + |B| - 3$. Then there exists $D \subseteq \mathbb{F}$ with $C \subseteq D$ and $|D| = |A| + |B| - 4$. Let

$$Q(x, y) = (x - y) \prod_{d \in D} (x + y - d).$$

As d varies over D, it varies over C as well. Taking $x = a, y = b$ with $a \neq b$ and $d = a + b$, we have $x + y - d = 0$.

Hence $Q(a, b) = 0$ for all $a \in A$, $b \in B$. The degree of $Q(x, y)$ is $|D| + 1$ and the coefficient $c_{Q,(i,|D|+1-i)}$ of $x^i y^{|D|+1-i}$ in $Q(x, y)$ arises in two ways:

(i) when x in $(x - y)$ is multiplied by $x^{i-1} y^{|D|+1-i} \in \prod_{d \in D} (x + y - d)$, and

(ii) when $-y$ in $(x - y)$ is multiplied by $x^i y^{|D|-i}$ in $\prod_{d \in D} (x + y - d)$.

Hence

$$c_{Q,(i,|D|+1-i)} = \binom{|D|}{i-1} - \binom{|D|}{i} = \binom{|D|}{i-1} \left[1 - \frac{|D| - i + 1}{i}\right].$$

This is zero if and only if $i = \dfrac{|D| + 1}{2}$ in $\mathbb{Z}_p$.

Therefore at least one of $c_{Q,(|A|-1,|B|-2)}$ and $c_{Q,(|A|-2,|B|-1)}$ is non-zero. By CNS, $\exists \alpha \in A$ and $\beta \in B$ with $Q(\alpha, \beta) \neq 0$. But this is a contradiction to the fact that $Q(x, y) = 0$ for every $x \in A$ and $y \in B$. This contradiction establishes the theorem in Case 2. ∎

Remark The proof given above is again based on Michałek [47] but it is not substantially different from the original proof of Alon [4].

2.3 Application of CNS to geometry

We now present an application of CNS to geometry due to Alon and Füredi [5].

Theorem 2.3.1 *Let $H_1, H_2, \ldots, H_m$ be a family of affine hyperplanes in $\mathbb{R}^n$ that cover all corners of the unit cube Q_n in $\mathbb{R}^n$ except one. Then $m \geq n$,*

Proof. We can assume, without loss of generality, that the uncovered vertex is the origin $(0, 0, \ldots 0)$. Let

$$a_{i1}x_1 + a_{i2}x_2 + \cdots + a_{in}x_n = b_i \qquad (2.2)$$

be the equation of the affine hyperplane H_i. Set

$$p_i(\mathbf{x}) = a_{i1}x_1 + a_{i2}x_2 + \cdots + a_{in}x_n. \qquad (2.3)$$

Here a_{ij} and b_i are real numbers. By our hypothesis, none of the planes H_i contains the origin. Hence $b_i \neq 0$ for each i.

Suppose the assertion is not true so that $m < n$. Now consider the polynomial

$$P(\mathbf{x}) = (-1)^{n+m+1}(b_1 b_2 \ldots b_m) \prod_{i=1}^{n}(x_i - 1) + \prod_{i=1}^{m}(p_i(\mathbf{x}) - b_i). \quad (2.4)$$

In the RHS of the above equation, the degree of the first term is n while the degree of the second term is m. As $m < n$, the degree of $P(\mathbf{x})$ is n and the coefficient of $(x_1 x_2 \ldots x_n)$ is $(-1)^{n+m+1}$ $b_1 b_2 \ldots b_m \neq 0$ since none of the b_i's is zero. Therefore by CNS (as the degree of each x_i is 1 in the leading monomial), if we take S_i to be the 2-element subset $\{0, 1\}$ of $\mathbb{R}$, there exists a point

$\mathbf{a} = (a_1, a_2, \ldots, a_n)$ with $a_i \in \{0, 1\}$ for each i and with $P(\mathbf{a}) \neq 0$. This point $\mathbf{a}$ is not the origin since

$$P(0) = (-1)^{n+m+1}(b_1 b_2 \ldots b_m)(-1)^n + (-b_1)(-b_2) \ldots (-b_m) = 0.$$

Hence $\mathbf{a}$ must be some other corner (as $a_i = 0$ or 1) of Q_n. But in this case, $p_i(\mathbf{a}) - b_i = 0$ for some i (as this corner is covered by some hyperplane H_i). The first term of $P(\mathbf{a})$ is zero (since some a_i must be 1) and once again we have $P(\mathbf{a}) = 0$. This contradiction proves that $m \geq n$. ∎

2.4   CNS and a subgraph problem

A well known conjecture of Berge and Sauer, proved by Tashkinov [59] and independently by Zhang [66], asserts that any simple 4-regular graph contains a 3-regular subgraph. This assertion is false for multi-graphs, i.e., graphs containing multiple edges. For instance, consider a 4-regular graph formed by duplicating the edges of an odd cycle. However, the existence of a 3-regular subgraph in any graph formed by adding a single edge (or increasing the multiplicity of an edge by 1) to any 4-regular multi-graph is a special case of the following more general result due to Alon, Friedland and Kalai [1], proved as an application of CNS.

Theorem 2.4.1 *For any prime p, and any loopless graph G, if the average degree of G is greater than $2p - 2$ and maximum degree is at most $2p - 1$, then G contains a p-regular subgraph.*

Proof. Assume that G is of order n and size m. Let $(a_{v,e})_{v \in V, e \in E}$ denote the $n \times m$ incidence matrix of G defined by

$$a_{ve} = \begin{cases} 1 & \text{if } v \text{ is incident to } e \\ 0 & \text{otherwise.} \end{cases}$$

Assume that the edges of G are $e_1, e_2, \ldots, e_m$ and associate with each edge e_i of G a variable x_i, and consider the polynomial

$$\begin{aligned}
F(\mathbf{x}) &= F(x_1, x_2, \ldots, x_m) \\
&= \prod_{v \in V} (1 - (a_{v,e_1}x_1 + a_{v,e_2}x_2 + \cdots + a_{v,e_m}x_m)^{p-1}) \\
&\quad - \prod_{i=1}^{m} (1 - x_i)
\end{aligned} \tag{2.5}$$

defined over the field $GF(p)$.

In (2.5), the first product has n terms while the second term has m terms. The degree of the second product is m while the degree of the first product is at most $n(p-1)$. By hypothesis, $2m/n > 2p-2$. So $m > n(p-1)$ and the second term in (2.5) is the dominant term with respect to the degree of $F(\mathbf{x})$. Moreover, the coefficient of $x_1 x_2 \ldots x_m$ in (2.5) is $-(-1)^m = (-1)^{m+1} \neq 0$. So by CNS, there are values $s_i \in \{0,1\} = S_i$ for each $i \in \{1, 2, \ldots, m\}$ such that $F(s_1, s_2, \ldots, s_m) \neq 0$. Note that not all s_i are zero as this would mean $F(s_1, s_2, \ldots, s_m) = 1-1 = 0$. Hence $\prod_{i=1}^{m}(1-s_i) = 0$ in (2.5). This implies that for any $v \in V$, the term $a_{v,e_1}s_1 + \cdots + a_{v,e_m}s_m = 0$. If not, this term is a non-zero element of $\mathbb{Z}_p$ and hence when raised to the power $p-1$ should give 1 in which case $F(s_1, s_2, \ldots, s_m) = 0$.

Let $E' = \{e_i : s_i = 1\}$ and $G' = (V, E')$. Then for each vertex v, the sum $a_{v,e_1}s_1 + \cdots + a_{v,e_m}s_m$ should be zero in $GF(p)$ for each $v \in V$. This means that for each vertex v, $d_{G'}(v) \equiv 0 \pmod{p}$. But as the maximum degree of G is at most $2p-1$, this degree is either 0 or p. Therefore the non-isolated vertices of G' induce a p-regular subgraph of G' and hence a p-regular subgraph of G. ∎

By taking $p = 3$ in Theorem 2.4.1, we obtain the following corollary:

Corollary 2.4.2 *Let G be a graph obtained from a 4-regular graph (multiple edges are allowed) by adding one edge. Then G contains a 3-regular subgraph.*

2.5 0–1 vectors in a hyperplane

A hyperplane in the n-dimensional vector space $GF(q)^n$ over the finite field $GF(q)$ is defined by an equation $a_1 x_1 + a_2 x_2 + \ldots + a_n x_n = 0$, where $\mathbf{a} = (a_1, a_2, \ldots, a_n)$ is the normal vector of H. In other words, $H = \mathbf{a}^\perp$ is the set of n-vectors over $GF(q)$ that are perpendicular to $\mathbf{a}$.

In this section, we present a result of Balandraud and Girard [9], which says that if p is a prime and $n > p$ and all the a_j's are non-zero, then the hyperplane $H = \mathbf{a}^\perp$ in $GF(p)^n$ is determined by the 0–1 vectors contained in H, that is, by the 0–1 vectors which are orthogonal to $\mathbf{a}$. The proof presented below is given by Schauz and Honold [54].

Theorem 2.5.1 *Assume that p is a prime and $n > p$ and H is a hyperplane in $GF(p)^n$ that does not contain any of the coordinate axes e_i, where $e_1 = (1, 0, \ldots, 0)$ etc. Then H has a basis consisting of 0–1 vectors.*

Proof. We need to show that the 0–1 vectors in $\mathbf{a}^\perp$ span a subspace of dimension $n - 1$ (namely span the whole hyperplane $\mathbf{a}^\perp$). This is true if and only if for any other hyperplane $\mathbf{b}^\perp$ in $GF(p)^n$, there is a 0–1 vector $(x_1, x_2, \ldots, x_n) \in \mathbf{a}^\perp$ which is not contained in $\mathbf{b}^\perp$.

Assume that $\mathbf{b} = (b_1, b_2, \ldots, b_n)$ and $\mathbf{b}^\perp \neq \mathbf{a}^\perp$, i.e., $\mathbf{b}$ is not a multiple of $\mathbf{a}$. Let $\lambda_i = \frac{b_i}{a_i}$. Since H does not contain any of the coordinate axes e_i, $a_i \neq 0$ for all i and λ_i is well-defined for all i. Let

$$P(\mathbf{x}) = \left(\sum_{i=1}^{n} \lambda_i x_i \right) \left(\left(\sum_{i=1}^{n} x_i \right)^{p-1} - 1 \right),$$

which is a polynomial of degree p in $GF(p)[x_1, x_2, \ldots, x_n]$. Since $\mathbf{b}^\perp \neq \mathbf{a}^\perp$, we may assume that $\lambda_1 \neq \lambda_2$. Let $\mathbf{t} = (t_1, t_2, \ldots, t_n)$, where $t_j = 1$ for $j \in \{2, 3, \ldots, p+1\}$, and $t_j = 0$ otherwise; and $\mathbf{t}' = (t'_1, t'_2, \ldots, t'_n)$, where $t'_j = 1$ for $j \in \{1, 3, \ldots, p+1\}$, and $t'_j = 0$ otherwise. As $n > p$, t_j, t'_j exist for $j \leq p+1$. Then for $j \in \{2, 3, \ldots, p+1\}$,

$$\lambda_j x_j \left(\left(\sum_{i=1}^{n} x_i \right)^{p-1} - 1 \right) = \lambda_j (p-1)! \mathbf{x}^{\mathbf{t}} + \text{other terms},$$

and for $j \in \{1, 3, \ldots, p+1\}$,

$$\lambda_j x_j \left(\left(\sum_{i=1}^{n} x_i \right)^{p-1} - 1 \right) = \lambda_j (p-1)! \mathbf{x}^{\mathbf{t}'} + \text{other terms}.$$

Hence

$$c_{P,\mathbf{t}} = (p-1)! \left(\left(\sum_{j=1}^{p+1} \lambda_j \right) - \lambda_1 \right), \text{ and}$$

$$c_{P,\mathbf{t'}} = (p-1)! \left(\left(\sum_{j=1}^{p+1} \lambda_j \right) - \lambda_2 \right).$$

So $c_{P,\mathbf{t}} - c_{P,\mathbf{t'}} \neq 0$, and hence at least one of $c_{P,\mathbf{t}}$ and $c_{P,\mathbf{t'}}$ is non-zero. By CNS, P has a non-zero point in $\{0, a_1\} \times \{0, a_2\} \times \ldots \times \{0, a_n\}$. In other words, there exists $(x_1, x_2, \ldots, x_n) \in \{0, 1\}^n$ such that $P(a_1 x_1, a_2 x_2, \ldots, a_n x_n) \neq 0$. This means that $\alpha = a_1 x_1 + a_2 x_2 + \cdots + a_n x_n = 0$ (since otherwise, $\alpha^{p-1} = 1$ in $GF(p)$, and and so the second factor in $P(a_1 x_1, a_2 x_2, \cdots, a_n x_n)$ would be zero) and $\beta = b_1 x_1 + b_2 x_2 + \cdots b_n x_n \neq 0$ (since otherwise the first factor in $P(a_1 x_1, \cdots, a_n x_n)$ would be zero). Hence $\mathbf{x} \in \mathbf{a}^\perp \backslash \mathbf{b}^\perp$. ∎

In the examples shown in this chapter, in order to solve our problem using CNS, we had to judiciously choose a suitable polynomial and invoke CNS. Needless to stress that it may not always be possible. On the other hand, there are also many combinatorial problems where the associated polynomials follow naturally from the definitions. The remainder of this book concentrates on the applications of CNS to graph colouring and graph labelling problems, where the associated polynomials are easily constructed. We shall concentrate on methods in proving that a certain monomial has non-zero coefficient in the expansion of a polynomial.

Chapter 3

Alon–Tarsi Theorem and Its Applications

3.1   Alon–Tarsi theorem

To apply CNS to solve a problem, one need to construct a polynomial so that the problem is reduced to deciding if the polynomial has a non-zero point in a certain grid. We then show that the coefficient of a certain monomial in the expansion of the polynomial is non-zero. In this monograph, we concentrate on the calculation of the coefficients of monomials, which is usually the difficult step in the application of CNS.

One of the most popular topics in graph theory is graph colouring. A widely studied variation of the conventional graph colouring, where one uses distinct colours for adjacent vertices, is the assignment of colours from out of a set of colours attached to each vertex of the graph. Such an assignment is known as list colouring. As mentioned by Alon [4], 'This variant received a considerable amount of attention that led to several facinating conjectures and results and its study combines interesting combinatorial techniques with powerful algebraic and probabilistic ideas'. This study is called 'choosability' in graphs and was initiated by Vizing [61], and independently by Erdős, Rubin and Taylor [23]. CNS is a useful tool in the study of choosability in graphs.

Definition 3.1.1 *Assume that $G = (V, E)$ is a graph. A* proper colouring *of G is a mapping ϕ which assigns to each vertex v a colour $\phi(v)$ from a certain colour set such that for each edge uv of G, $\phi(u) \neq \phi(v)$. A k-colouring of G is a proper colouring of G with colour set $\{1, 2, \ldots, k\}$. We say that G is k-colourable if there is a proper k-colouring of G. The* chromatic number $\chi(G)$ *of G is defined as*

$$\chi(G) = \min\{k : G \text{ is } k\text{-colourable}\}.$$

Definition 3.1.2 *Assume that G is a graph and $f : V(G) \to \mathbb{Z}^+$ is a function which assigns a positive integer $f(v)$ to each vertex $v \in V(G)$. An f-list assignment of G is a mapping L which assigns to each vertex v of G a list $L(v)$ of $f(v)$ elements of a field F as permissible colours. If $f(v) = k$ for all $v \in V(G)$, then the f-list assignment of G is called a k-list assignment of G. An L-colouring of G is a proper colouring ϕ of G such that $\phi(v) \in L(v)$ for each vertex v. We say that G is L-colourable if there is an L-colouring of G, and we say that G is f-choosable (respectively, k-choosable) if G is L-colourable for each f-list assignment (respectively, k-list assignment) L of G. The choice number $ch(G)$ of G is then defined as*

$$ch(G) = \min\{k : G \text{ is } k\text{-choosable}\}.$$

It follows from the definition that for any graph G, $\chi(G) \leq ch(G)$. On the other hand, the difference $ch(G) - \chi(G)$ can be arbitrarily large (cf. Theorem 1.0.15).

Definition 3.1.3 *Assume that G is an undirected graph whose vertices are linearly ordered and $\mathbb{F}$ is a field. We associate to each vertex v of G a variable x_v. The graph polynomial $\in \mathbb{F}[\mathbf{x}]$ of G is defined as*

$$f_G(\mathbf{x}) = \prod_{uv \in E(G), u < v} (x_u - x_v).$$

In the definition of $f_G(\mathbf{x})$, we used a linear order of the vertices of G. The order is just used to determine, for each edge uv of G, whether the linear term in the product is $(x_u - x_v)$ or $(x_v - x_u)$. So if $g_G(\mathbf{x})$ is defined by using a different linear order, then it is easy to see that $g_G(\mathbf{x}) = \pm f_G(\mathbf{x})$. Instead of using a linear ordering of the vertices of G, we may use an orientation D of G to define the polynomial as

$$f_D(\mathbf{x}) = \prod_{(u,v) \in E(D)} (x_u - x_v).$$

For different orientations D, D' of G, we have $f_D(\mathbf{x}) = \pm f_{D'}(\mathbf{x})$. We shall be mainly interested in whether the coefficient of a certain monomial $\mathbf{x}^\mathbf{t}$ in the expansion of $f(\mathbf{x})$ is non-zero. Thus, for this purpose, it is immaterial if we choose $f(\mathbf{x})$ or $-f(\mathbf{x})$. So in defining the polynomial $f_G(\mathbf{x})$, the orientation D (or the linear ordering of the vertices of G) is not important, and we may write $f_G(\mathbf{x})$ for $f_D(\mathbf{x})$. Nevertheless, for the convenience of discussion, sometimes we shall choose the orientation D carefully.

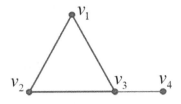

FIGURE 3.1: Example: A graph G with $f_G(\mathbf{x}) = (x_1 - x_2)(x_1 - x_3)(x_2 - x_3)(x_3 - x_4)$.

Assume that $\phi : V(G) \to \mathbb{F}$ is a mapping which assigns to each vertex v an element $\phi(v) \in \mathbb{F}$. We denote by $f_G(\phi)$ the evaluation of f_G at $x_v = \phi(v)$ for $v \in V(G)$. It is obvious that ϕ is a proper colouring of G if and only if $f_G(\phi) \neq 0$. Thus to prove that G is k-choosable, it suffices to show that f_G has a non-vanishing monomial $\prod_{v \in V(G)} x_v^{t_v}$ such that $t_v \leq k - 1$ for $v \in V(G)$. For, in this case, we can take $|S_v| = k$ for each v and invoke CNS. This motivates the definition of Alon–Tarsi number given below. Observe that the polynomial f_G is a homogeneous polynomial of degree $m = E(G)$. Thus each non-vanishing monomial of f_G is of highest degree.

Definition 3.1.4 *We denote by $\mathbb{N}_0^G$ the set of all mappings $f :$ $V(G) \to \{0, 1, 2, \ldots\}$. For $f, g \in \mathbb{N}_0^G$, we write $f \leq g$ if $f(v) \leq g(v)$ for every vertex v. Let*

$$\mathbf{x}^g := \prod_{v \in V(G)} x_v^{g(v)}.$$

For convenience, we denote by $c_{G,g}$ the coefficient of $\mathbf{x}^g$ in the expansion of f_G, i.e., $c_{G,g} = c_{f_G,g}$. If $g(v) = d$ for some constant d, then we write $c_{G,d}$ for $c_{G,g}$.

Definition 3.1.5 *Assume that G is a graph and $f \in \mathbb{N}_0^G$. We say that G is Alon–Tarsi f-choosable, or f-AT for short, if f_G has a non-vanishing monomial $\mathbf{x}^g$ with $g(v) \leq f(v) - 1$ for $v \in V(G)$. We say G is k-AT if G is f-AT for the constant function $f(v) = k$ for every $v \in V(G)$. The Alon–Tarsi number $AT(G)$ of G is defined as*

$$AT(G) = \min\{k : G \text{ is } k\text{-AT}\}.$$

For example, the graph G in Figure 3.1 has $AT(G) = 3$, as $x_2^2 x_1 x_3$ is a non-vanishing monomial of $f_G(\mathbf{x})$, and each non-vanishing monomial contains a variable to the power of 2 or 3 (the monomial $x_1 x_2 x_3 x_4$ has coefficient 0).

It follows from CNS that for any graph G,

$$ch(G) \leq AT(G).$$

On the other hand, the difference $AT(G) - ch(G)$ can be arbitrarily large.

Observation 1 *For* $G = K_{n,n}$, $AT(G) \geq n/2$ *and* $ch(G) \leq \log_2 n + 2$.

Proof. Since f_G has degree $|E(G)| = n^2$, it follows that for any non-vanishing monomial $\mathbf{x}^g$ of f_G, $g(v) \geq n^2/|V(G)| = n/2$ for some v. Hence $AT(G) \geq n/2$.

Assume that $k = \lfloor \log_2 n \rfloor + 2$ and L is a k-list assignment of G. Without loss of generality, assume that $\bigcup_{v \in V(G)} L(v) = \{1, 2, \ldots, q\}$. Let $X_i = \{v : i \in L(v)\}$. Assume that the two partite sets of G are A, B. We assign weight $w(v) = 1$ to each vertex v of G and for each subset X of $V(G)$, let $w(X) = \sum_{v \in X} w(v)$ be the weight of X.

We colour the vertices of G in q steps,

The weights of vertices in G will be updated in each step, and we always use $w(v)$ to denote the current weight of v. Let U_i be the set of uncoloured vertices before the ith step. So $U_1 = V(G)$. In the ith step, if $w(X_i \cap U_i \cap A) \geq w(X_i \cap U_i \cap B)$, then we colour vertices in $X_i \cap U_i \cap A$ with colour i and double the weight of each vertex in $X_i \cap U_i \cap B$. Otherwise, we colour vertices in $X_i \cap U_i \cap B$ with colour i and double the weight of each vertex in $X_i \cap U_i \cap A$.

It follows from the procedure above that the total weight of uncoloured vertices is never increased, and the weight of an uncoloured vertex $v \in U_i$ has weight $w(v) = 2^{m_i(v)}$, where $m_i(v) = |L(v) \cap \{1, 2, \ldots, i-1\}|$. As the total weight of $V(K_{n,n})$ is $2n$, we conclude that each uncoloured vertex v has weight at most $2n$. Since $m_{q+1}(v) = k$, after the qth step, if v is still uncoloured, then $w(v) = 2^k > 2n$, which is a contradiction. Thus after the qth step, all vertices are coloured and we obtain an L-colouring of G. This proves that $ch(G) \leq \log_2 n + 2$. ∎

It is known [63] that for any positive integer n, $ch(G) \leq \log_2 n - (\frac{1}{2} + o(1)) \log_2 \log_2 n$. It was conjectured by Erdős and Lovász (cf. [63]) that $ch(G) = \Omega(\log_2 n - (1 + o(1)) \log_2 \log_2 n)$ and the conjecture remains open.

Definition 3.1.6 *For* $f \in \mathbb{N}_0^G$ *and* $u \in V(G)$ *for which* $f(u) > 0$, *define* $\hat{f}_u \in \mathbb{N}_0^G$ *by setting*

$$\hat{f}_u(v) = \begin{cases} f(v) - 1, & \text{if } v = u, \\ f(v), & \text{otherwise.} \end{cases}$$

Lemma 3.1.7 *If* H *is a subgraph of* G, $f \in \mathbb{N}_0^G$ *and* G *is* f-AT, *then* H *is* f-AT. *In particular,* $AT(H) \leq AT(G)$.

Proof. It suffices to prove that if H is obtained from G by deleting a single edge $e = uv$, then H is f-AT. In this case,

$$f_G = (x_u - x_v)f_H = x_u f_H - x_v f_H.$$

Assume that $\mathbf{x}^g$ is a non-vanishing monomial in f_G. Then at least one of $\mathbf{x}^{\hat{g}_u}, \mathbf{x}^{\hat{g}_v}$ is a non-vanishing monomial in f_H, for otherwise $\mathbf{x}^g$ is a vanishing monomial in both $x_u f_H$ and $x_v f_H$, and hence $\mathbf{x}^g$ cannot be a non-vanishing monomial of f_G. Therefore H is f-AT. ∎

Recall that a digraph D' is called Eulerian if for every vertex v, $d_{D'}^+(v) = d_{D'}^-(v)$, $\mathcal{E}(D)$ the family of spanning Eulerian subdigraphs of D, and

$$\begin{aligned} \mathcal{E}_e(D) &= \{D' \in \mathcal{E}(D) : |A(D')| \equiv 0 \pmod 2\}, \\ \mathcal{E}_o(D) &= \{D' \in \mathcal{E}(D) : |A(D')| \equiv 1 \pmod 2\}, \\ \text{diff}(D) &= |\mathcal{E}_e(D)| - |\mathcal{E}_o(D)|. \end{aligned}$$

Definition 3.1.8 *We say that* D *is* Alon–Tarsi *if* $\text{diff}(D) \neq 0$. *If an orientation* D *of* G *yields an Alon–Tarsi digraph, then we say* D *is an Alon–Tarsi orientation (or an AT-orientation, for short) of* G.

Alon and Tarsi [3] proved the following theorem. Its significance rests in the fact that it transforms the computation of the Alon–Tarsi number of a graph G from algebraic manipulations to the analysis of the structural properties of G. (See, for instance, Section 3.4).

Theorem 3.1.9 (Alon–Tarsi Theorem) *A graph G is* f-AT *if and only if G has an AT-orientation* D *with* $d_D^+(v) \leq f(v) - 1$ *for each vertex* v *of G. Therefore*

$$AT(G) = \min\{k : G \text{ has an AT-orientation } D \text{ with } \Delta_D^+ = k - 1\}.$$

Proof. In calculating the expansion of $f_G(\mathbf{x}) = \prod_{uv \in E(G), u < v}(x_u - x_v)$, we first do $2^{|E(G)|}$ multiplications, where each multiplication is done by choosing, for each edge $uv \in E(G)$ with $u < v$, either x_u or $-x_v$ in the multiplication. We use an orientation D of G to indicate for each edge, whether x_u or $-x_v$ is chosen. Assume that D is an orientation of G. For an arc $e = (u, v)$ of D, we define the weight $w(e)$ to be $w(e) = x_u$ if $u < v$ and $w(e) = -x_u$ if $u > v$, and let $w(D) = \prod_{e \in A(D)} w(e)$. Then

$$f_G(\mathbf{x}) = \sum_D w(D), \qquad (3.1)$$

where the summation ranges over all the $2^{|E(G)|}$ orientations D of G.

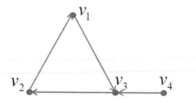

FIGURE 3.2: Example: an orientation D with $|\mathcal{E}_e(D)| = |\mathcal{E}_o(D)| = 1$.

For instance, for the oriented graph D of G of Figure 3.2, $w(D) = \prod_{e \in A(D)} w(e) = w(e_1)w(e_2)w(e_3)w(e_4) = x_1(-x_2)(-x_3)(-x_4) = -x_1x_2x_3x_4$.

Let $m(D)$ be the number of arcs of D in 'reverse' directions, i.e.,

$$m(D) = |\{(u, v) \in A(D) : u > v\}|.$$

It follows from the definition that

$$w(D) = (-1)^{m(D)} \prod_{v \in V(G)} x_v^{d_D^+(v)}.$$

We say D is *even* (or *odd*) if $m(D)$ is even (or odd). For $g \in \mathbb{N}_0^G$, let

$$\mathcal{E}(G; g) =$$
$$\{D : D \text{ is an orientation of } G \text{ for which } d_D^+(v) = g(v) \text{ for all } v \in V(G)\}$$

Let

$$
\begin{aligned}
\mathcal{E}_e(G;g) &= \{D \in \mathcal{E}(G;g) : D \text{ is an even orientation of } G\}, \\
\mathcal{E}_o(G;g) &= \{D \in \mathcal{E}(G;g) : D \text{ is an odd orientation of } G\}.
\end{aligned}
$$

Then

$$
f_G(\mathbf{x}) = \sum_{g \in \mathbb{N}_0^G} \left(|\mathcal{E}_e(G;g)| - |\mathcal{E}_o(G;g)| \right) \mathbf{x}^g.
$$

Assume that $g \in \mathbb{N}_0^G$ and $D \in \mathcal{E}(G;g)$. For any $D' \in \mathcal{E}(G;g)$, let $D \oplus D'$ be the set of arcs in D whose directions are reversed in D', i.e.,

$$
D \oplus D' = \{(u,v) \in A(D) : (v,u) \in A(D')\}.
$$

For each vertex v, since $d_D^+(v) = d_{D'}^+(v)$, the number of outgoing arcs of D that got changed into incoming arcs of D' equals the number of incoming arcs of D that have been changed into outgoing arcs of D'. Hence $D \oplus D'$ is an Eulerian subdigraph of D. i.e., $D \oplus D' \in \mathcal{E}(D)$. Conversely, if $H \in \mathcal{E}(D)$ and D' is obtained from D by reversing the orientation of the arcs in H, then D' and D have the same out-degree sequence.

Moreover, $D \oplus D' \in \mathcal{E}_e(D)$ if D and D' have the same parity (that is, either both D and D' are even or both D and D' are odd), and $D \oplus D' \in \mathcal{E}_o(D)$ if D and D' have opposite parities. The mapping $D' \to D \oplus D'$ is a bijection between $\mathcal{D}_e(G;g) \cup \mathcal{D}_o(G;g)$ and the set of Eulerian subdigraphs of D. When D is even, it maps even orientations to even (Eulerian) subdigraphs and odd orientations to odd subdigraphs. Otherwise it maps even orientations to odd subdigraphs and odd orientations to even subdigraphs. In both the cases,

$$
|\mathcal{D}_e(G;g)| - |\mathcal{D}_o(G;g)| = \pm\mathrm{diff}(D).
$$

So the coefficient of the monomial $\prod_{i=1}^n x_i^{d_i}$ in the expansion of f_G is $\pm\mathrm{diff}(D)$. In particular, $\prod_{i=1}^n x_i^{d_i}$ is a non-vanishing monomial of f_G if and only if D is an AT-orientation of G.

Hence

$$
AT(G) = \min\{k : G \text{ has an AT-orientation } D \text{ with } \Delta_D^+ = k - 1\}.
$$

This completes the proof of Theorem 3.1.9. ∎

Example 3.1.10 *The orientation in Figure 3.2 is the only orientation of the underlying graph G with maximum out-degree 1. As this orientation is not an AT-orientation, we conclude that $AT(G) \geq 3$. On the other hand, by reversing the direction of one edge in the triangle, the resulting orientation is an AT-orientation. So $AT(G) = 3$.*

Lemma 3.1.11 *Let $d_1 \geq d_2 \geq \cdots \geq d_n$ be the out-degree sequence of a digraph D, satisfying $|\mathcal{E}_e(D)| \neq |\mathcal{E}_o(D)|$. Then for every $k, n \geq k \geq 0$, G has an independent set of size at least $\left\lceil \dfrac{(n-k)}{(d_{k+1}+1)} \right\rceil$.*

Proof. Relabel the vertices of G, if necessary, so that $d^+(v_i) = d_i$. By Theorem 3.1.9, there is a legal vertex colouring $c : V \to \mathbb{Z}$ such that $1 \leq c(v_i) \leq d_i + 1$ for each $i, 1 \leq i \leq n$. Let $k + 1 \leq j \leq n$. Then $1 \leq c(v_j) \leq d_j + 1 \leq d_{k+1} + 1$. Hence to colour the $n - k$ vertices $v_{k+1}, v_{k+2}, \ldots, v_n$, at most $d_{k+1}+1$ colours have been used. So at least one colour class has size at least $\left\lceil \dfrac{(n-k)}{(d_{k+1}+1)} \right\rceil$. ∎

Assume that D is a digraph. For a subset A of arcs of D, denote by A^R the set of arcs obtained by reversing the orientation of the arcs in A. For convenience, a digraph is usually viewed as a set of arcs. Thus $(D - A) \cup A^R$ denotes the digraph obtained from D by reversing the direction of the arcs in A.

Lemma 3.1.12 *Assume that D is a digraph and C is a directed cycle in D. Let $D' = (D - C) \cup C^R$. For an Eulerian subdigraph H of D, let $\phi(H) = (H - C) \cup (C - H)^R$. Then ϕ is a one-to-one correspondence between $\mathcal{E}(D)$ and $\mathcal{E}(D')$. In particular, $|\mathcal{E}(D)| = |\mathcal{E}(D')|$.*

Proof. First we show that $\phi(H)$ is an Eulerian subdigraph of D'. It follows from the definition that $\phi(H)$ is a subdigraph of D'. Assume that $x \in V(D)$. We calculate the changes in the in-degree and out-degree of x as H is changed to $\phi(H)$. If $x \notin V(C)$, then the in-degree and out-degree of x are not changed. Assume that $x \in V(C)$. If both the arcs of C incident to x are contained in H, then both the in-degree and the out-degree of x decrease by 1. If both the arcs of C incident to x are not contained in H, then both the in-degree and the out-degree of x increase by 1. If exactly one arc incident to x is contained in H, then the in-degree and out-degree x are not changed. Thus $\phi(H)$ is an Eulerian subdigraph of D'.

The same argument shows that for any Eulerian subdigraph H of D', $\psi(H) = (H - C^R) \cup (C^R - H)^R$ is an Eulerian subdigraph of D. Moreover, for any $H \in \mathcal{E}(D)$, $\psi(\phi(H)) = H$. Thus ϕ is a one-to-one correspondence between $\mathcal{E}(D)$ and $\mathcal{E}(D')$. ∎

Corollary 3.1.13 *For any two orientations D_1, D_2 of a graph G with the same out-degree sequence, $|\mathcal{E}(D_1)| = |\mathcal{E}(D_2)|$.*

Proof. The subdigraph $D_1 \oplus D_2$ is an Eulerian subdigraph of D_1 and hence it is the arc disjoint union of directed cycles. So D_2 can be obtained from D_1 by recursively reversing the directions of directed cycles, and hence $|\mathcal{E}(D_1)| = |\mathcal{E}(D_2)|$. ∎

Definition 3.1.14 *A digraph D is* strongly connected *if for any two vertices u and v, there is a directed path from u to v.*

Definition 3.1.15 *For a digraph D, a* strongly connected component *of D is a maximal subdigraph of D which is strongly connected.*

Observe that for any digraph D, there is a partial ordering $\leq$ of the strongly connected components of D such that for any two strongly connected components A and B of D, if $A < B$, then all the arcs between A and B (if any) are from A to B.

Lemma 3.1.16 *Assume that D is a digraph and $V(D) = X_1 \dot\cup X_2$. For $i = 1, 2$, let $D_i = D[X_i]$ be the subdigraph of D induced by X_i. If all the arcs between X_1 and X_2 are from X_1 to X_2, then D is Alon–Tarsi if and only if D_1, D_2 are both Alon–Tarsi.*

Proof. Since any arc of an Eulerian digraph is contained in a directed cycle, each Eulerian subdigraph H of D is the arc disjoint union of H_1 and H_2, where H_i is an Eulerian subdigraph of D_i. Now H is even if and only if H_1, H_2 have the same parity. Therefore

$$\begin{aligned}
|\mathcal{E}_e(D)| &= |\mathcal{E}_e(D_1)| \times |\mathcal{E}_e(D_2)| + |\mathcal{E}_o(D_1)| \times |\mathcal{E}_o(D_2)|, \\
|\mathcal{E}_o(D)| &= |\mathcal{E}_e(D_1)| \times |\mathcal{E}_o(D_2)| + |\mathcal{E}_o(D_1)| \times |\mathcal{E}_e(D_2)|.
\end{aligned}$$

So $\mathrm{diff}(D) = \mathrm{diff}(D_1) \times \mathrm{diff}(D_2)$ and D is AT if and only if both D_1, D_2 are AT. ∎

Corollary 3.1.17 *A digraph is Alon–Tarsi if and only if each of its strongly connected components is Alon–Tarsi.*

3.2 Bipartite graphs and acyclic orientations

In general, it is difficult to determine if an orientation D of a graph G is AT. However, there are some cases where this problem is easy. Observe that every digraph D has at least one even Eulerian subdigraph, namely, the empty subdigraph, and every Eulerian digraph can be decomposed into directed cycles. So if D has no odd directed cycle, then D has no odd Eulerian subdigraph and hence D is AT. Thus we have the following result.

Lemma 3.2.1 *If G is a bipartite graph, then every orientation D of G is an AT-orientation. For any graph G, any acyclic orientation D of G is an AT-orientation of G.*

Corollary 3.2.2 *If G is a bipartite graph, then $AT(G) = \max_{H \subset G} \left\lceil \frac{|E(H)|}{|V(H)|} \right\rceil + 1$.*

Proof. Assume that $k = \max_{H \subset G} \left\lceil \frac{|E(H)|}{|V(H)|} \right\rceil$.

For any orientation D of G and for any subgraph H of G, the restriction of D to H has maximum out-degree at least $\left\lceil \frac{|E(H)|}{|V(H)|} \right\rceil$. Hence $AT(G) \geq k + 1$.

We now show that there is an orientation D of G for which $\Delta_D^+ \leq k$. We construct a bipartite graph Q with partite sets $E(G)$ and $V(G) \times [k]$, where $[k] = \{1, 2, \ldots, k\}$, and $(v, j) \in V(G) \times [k]$ is adjacent to $e \in E(G)$ if v is an end vertex of e. For any subset X of $E(G)$, let Y be the set of vertices of G each of which is incident to at least one edge of X. Then $N_Q(X) = Y \times [k]$. Since $|X| \leq k|Y|$, we conclude that $|X| \leq |N_Q(X)|$. By Hall's Theorem, Q has a matching M that saturates every $e \in E$. We orient the edge $e = uv \in E(G)$ as (u, v) if and only if $(e, (u, j)) \in M$ for some j. Then the resulting orientation D has $\Delta_D^+ \leq k$. By Theorem 3.1.9, $AT(G) \leq k + 1$. Hence $AT(G) = k + 1$. ∎

Corollary 3.2.3 *Bipartite planar graphs G have $AT(G) \leq 3$ and hence are 3-choosable.*

Proof. It follows from Euler's formula (Theorem 1.0.10) that for any subgraph H of G, $|E(H)|/|V(H)| \leq 2$. ∎

The *colouring number* of a graph G is defined as

$$\mathrm{col}(G) = 1 + \max\{\delta(H) : H \text{ is a subgraph of } G\},$$

where $\delta(H)$ stands for the minimum degree of H.

Alternately, $\mathrm{col}(G)$ is the minimum integer k such that there is a linear ordering $<$ of the vertices of G so that each vertex v has at most $k-1$ neighbours u with $u < v$. By orienting each edge uv as (u, v) if $u > v$, we obtain an acyclic orientation D of G with $\delta_D^+ = \mathrm{col}(G) - 1$. Hence we have the following corollary.

Corollary 3.2.4 *For any graph G, $AT(G) \leq \mathrm{col}(G)$.*

3.3 The Cartesian product of a path and an odd cycle

Another simple sufficient condition for an orientation D to be Alon–Tarsi is that $|\mathcal{E}(D)|$ is odd. For, if $|\mathcal{E}_e(D)| = |\mathcal{E}_o(D)|$, then $|\mathcal{E}(D)| = |\mathcal{E}_e(D)| + |\mathcal{E}_o(D)|$ is even. Hence if $|\mathcal{E}(D)|$ is odd, then $|\mathcal{E}_e(D)| \neq |\mathcal{E}_o(D)|$ and D is Alon–Tarsi. In this section, this idea is used to prove that the Cartesian product of a path and an odd cycle has Alon–Tarsi number 3.

Definition 3.3.1 *Assume that $G = (V, E)$ and $G' = (V', E')$ are graphs. The Cartesian product $G \square G'$ is the graph with vertex set $V \times V'$, in which (x, x') is adjacent to (y, y') if and only if either $x = x'$ and $yy' \in E'$ or $y = y'$ and $xx' \in E$. Figure 3.3 is the graph $P_3 \square C_5$.*

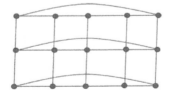

FIGURE 3.3: The graph $P_3 \square C_5$.

Theorem 3.3.2 *[34] If G is the Cartesian product $C_{2k+1} \square P_n$ of an odd cycle and a path, then $AT(G) = 3$.*

Proof. Since G has chromatic number 3, we know that $AT(G) \geq 3$.

It remains to show that $AT(G) \leq 3$. If $n = 1$, then the result is obvious. We assume that $n \geq 2$.

Assume that the vertices of G are $\{(v_i, w_j) : 1 \leq i \leq 2k+1, 1 \leq j \leq n\}$, where for each j, the set $X_j = \{(v_i, w_j) : 1 \leq i \leq 2k+1\}$ induces an odd cycle of length $2k + 1$, and for each i, the set $Y_i = \{(v_i, w_j) : 1 \leq j \leq n\}$ induces a path on n vertices. Let D be the orientation of G in which the edges of G are oriented in such a way that for each j, $G[X_j]$ is a directed cycle, and for each i, $G[Y_i]$ is a directed path from (v_i, w_1) to (v_i, w_n) [See Figure 3.4].

Let D^* be obtained from D by adding an arc e^* from (v_2, w_n) to (v_1, w_1). It is easy to check that D^* has maximum out-degree 2. Now we show that D^* has an odd number of Eulerian subdigraphs.

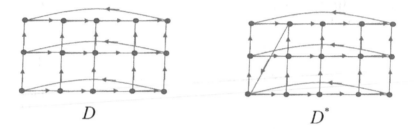

D D^*

FIGURE 3.4: The digraphs D and D^* for $n = 3$ and $k = 2$.

Note that the only directed cycles in D are those directed cycles induced by X_j for $j = 1, 2, \ldots, n$. All the directed cycles in D^* not contained in D contain the arc e^*.

Let $\mathcal{A}$ be the family of Eulerian sub-digraphs Q of D^* such that for $1 \leq j \leq n$, either all the arcs of $D(X_j)$ are contained in Q or disjoint from Q. That is,

$$\mathcal{A} = \{Q \in \mathcal{E}(D^*) : \exists j, E(D[X_j]) \subseteq E(Q) \text{ or } E(D[X_j]) \cap E(Q) = \emptyset\}.$$

For each Eulerian subdigraph Q in $\mathcal{A}$, let Q' be obtained from Q by deleting all the arcs of those directed cycles $D[X_j]$ contained in Q, and adding the arcs of those directed cycles $D[X_j]$ that are arc disjoint from Q. By definition, Q and Q' are distinct members of $\mathcal{A}$, and each member of $\mathcal{A}$ is contained in exactly one such pair. Therefore $|\mathcal{A}|$ is even.

Let $\mathcal{B}$ be the set of all the other Eulerian subdigraphs of D^*. Since each Eulerian subdigraph can be decomposed into an arc-disjoint union of directed cycles (and the only directed cycles in D are $D[X_j]$, for $1 \leq j \leq n$), we conclude that each member of $\mathcal{B}$ is a directed cycle containing e^* and contains at least one arc and at most $2k$ arcs from each directed cycle $D[X_j]$. To construct such a directed cycle C, we start from (v_1, w_1), traverse at least one arc but at most $2k$ arcs from $D[X_1]$, then go to X_2 and traverse at least one arc but at most $2k$ arcs from $D[X_2]$, and then continue in the same fashion to $X_3, X_4, \ldots, X_n$. Such a directed cycle is uniquely determined by the positions in which it goes from X_i to X_{i+1} for $i = 1, 2, \ldots, n-1$. Since the last arc of the directed cycle is from (v_2, w_n) to (v_1, w_1), for the directed cycle to contain at least one and at most $2k$ arcs of $D[X_n]$, it is required that the arc from X_{n-1} to X_n is not the arc joining (v_2, w_{n-1}) to (v_2, w_n).

Let S_n, S'_n and S''_n be sets of integer sequences defined as follows:

$$
\begin{aligned}
S_n &= \{(i_1, i_2, \ldots, i_{n-1}) : 1 \le i_j \le 2k+1, i_1 \ne 1, i_{j+1} \ne i_j\}, \\
S'_n &= \{(i_1, i_2, \ldots, i_{n-1}) \in S_n : i_{n-1} \ne 2\}, \\
S''_n &= \{(i_1, i_2, \ldots, i_{n-1}) \in S_n : i_{n-1} = 2\}.
\end{aligned}
$$

By the argument above, $|\mathcal{B}| = |S'_n|$. To form a sequence $(i_1, i_2, \ldots, i_{n-1})$ in S_n, each i_j has $2k$ choices. So $|S_n| = (2k)^{n-1}$. It is obvious that $|S'_2| = 2k - 1$ and $|S''_2| = 1$ are odd. Each sequence in S''_{n-1} can be extended to a sequence in S'_n in $2k$ different ways (i_{n-1} can be any index distinct from i_{n-2}), and each sequence in S'_{n-1} can be extended to a sequence in S'_n in $2k - 1$ different ways (i_{n-1} can be any index distinct from both i_{n-2} and 2). So $|S'_n| = |S''_{n-1}|2k + |S'_{n-1}|(2k - 1)$. As $|S''_n| = |S_n| - |S'_n|$, S'_n and S''_n have the same parity. Thus by induction, $|S'_n|$ and $|S''_n|$ are both odd for all n. So $|\mathcal{B}|$ is odd, and hence D^* has an odd number of Eulerian subdigraphs.

Thus D^* is an AT-orientation of $G^* = G + e$, where e is an edge connecting (v_1, w_2) to (v_2, w_n). Hence $AT(G^*) \le 3$, and by Lemma 3.1.7, $AT(G) \le 3$. ∎

Corollary 3.3.3 *If $G = C_{2k+1} \Box P_n$ for some positive integers k, n, then $\chi(G) = ch(G) = AT(G) = 3$.*

If $G = C_{2k} \Box P_n$ and $n \ge 2$, then G is a bipartite graph with $1 < |E(G)|/|V(G)| < 2$. By Corollary 3.2.2, $AT(G) = 3$. So the Cartesian product of any cycle with a path with at least two vertices has Alon–Tarsi number 3.

It is natural to ask as to what is the Alon–Tarsi number of the Cartesian product $C_n \Box C_m$ of two cycles. If both n and m are even, then $C_n \square C_m$ is bipartite, and it follows from Corollary 3.2.2 that $C_n \square C_m$ has Alon–Tarsi number 3. When at least one of n, m is odd, this question will be answered in Section 4.6.

3.4   A solution to a problem of Erdős

Similar to the idea in the previous section, if $|\mathcal{E}_e(D)| = |\mathcal{E}_o(D)|$ is even, then $|\mathcal{E}(D)| \equiv 0 \pmod 4$. Thus if $|\mathcal{E}(D)| \equiv 2 \pmod 4$ and $\mathcal{E}_e(D)$ is even, then D is an Alon–Tarsi orientation. In this section, this technique is used to prove a family of 4-regular graphs has Alon–Tarsi number 3.

Assume that D is an Eulerian digraph. For any Eulerian subdigraph H of D, its complement $D \backslash H$ is also an Eulerian digraph. (Recall that as per our definition, an Eulerian digraph need not be connected). If D has an odd number of arcs, then H and $D \backslash H$ have opposite parity. Hence $|\mathcal{E}_e(D)| = |\mathcal{E}_o(D)|$. So D is not an AT-orientation. On the other hand, if D has an even number of arcs, then H and $D \backslash H$ have the same parity. Thus $|\mathcal{E}_e(D)|$ and $|\mathcal{E}_o(D)|$ are even. Therefore, if $|\mathcal{E}(D)| \equiv 2 \pmod 4$, then $|\mathcal{E}_e(D)| \neq |\mathcal{E}_o(D)|$ and D is an Alon–Tarsi orientation.

Assume that G is a 4-regular graph on $3n$ vertices whose edge set can be decomposed into a Hamiltonian cycle and n pairwise vertex-disjoint triangles. (See Figure 3.5 for an example of such a graph). Then an independent set of G contains at most one vertex from each triangle, and hence has size at most n. Du and Hsu conjectured in 1986 that G has independence number n. In 1990, Erdős asked whether G has chromatic number 3. This problem, often referred to as the 'cycle plus triangles problem of Erdős', was solved in the affirmative by Fleischner and Stiebitz in 1992 [25]. Indeed, Fleischner and Stiebitz proved the following stronger result.

Theorem 3.4.1 *Assume that G is a 4-regular graph on $3n$ vertices whose edge set can be decomposed into a Hamiltonian cycle and n pairwise vertex disjoint triangles. Then $AT(G) = 3$.*

To prove Theorem 3.4.1, we need to find an AT-orientaiton D of G in which each vertex has out-degree at most 2. Since $\sum_{x \in V(D)} d_D^+(x) = |A(D)| = 2|V(D)|$, we conclude that $d_D^+(x) =$

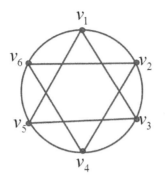

FIGURE 3.5: An example of a graph G of Theorem 3.4.1.

$d_D^-(x) = 2$ for each vertex x. So D is an Eulerian digraph. Theorem 3.4.1 follows from the next lemma.

Lemma 3.4.2 *Assume that D is an Eulerian digraph whose edge set can be decomposed into a directed Hamiltonian cycle C and $n \geq 0$ pairwise vertex-disjoint directed triangles. Then $|\mathcal{E}(D)| \equiv 2$ (mod 4).*

Proof. As the n triangles are vertex disjoint, $|V(D)| \geq 3n$. We observe that the number of vertices of D can be greater than n. Thus some vertices of D may not be contained in any of the n triangles, and hence has in-degree and out-degree both equal to 1.

We prove the lemma by induction on n. If $n = 0$, then D is a directed cycle and $|\mathcal{E}(D)| = 2$ (the only Eulerian subdigraphs of D are the empty digraph and D itself).

Assume that $n \geq 1$ and the Lemma holds if there are at most $n - 1$ triangles.

Let T be one of the triangles as indicated in Figure 3.6, where $V(T) = \{x_1, x_2, x_3\}$ and the arcs of T are $a_1 = (x_3, x_2), a_2 = (x_1, x_3), a_3 = (x_2, x_1)$.

Now an Eulerian subdigraph of D may contain one, two, all or none of the arcs a_1, a_2, a_3. Recall that for arcs $z_1, z_2, \ldots, z_q, y_1, y_2, \ldots, y_p$, $\mathcal{E}(D, z_1, z_2, \ldots, z_q, \overline{y}_1, \overline{y}_2, \ldots, \overline{y}_p)$ is the family of Eulerian subdigraphs H of D which contain $z_1, z_2, \ldots, z_q$ but which do not contain any of $y_1, y_2, \ldots, y_p$.

Hence

$$\mathcal{E}(D) = \mathcal{E}^*(D) \dot\cup \mathcal{E}(D, a_1, a_2, a_3) \dot\cup \mathcal{E}(D, \overline{a}_1, \overline{a}_2, \overline{a}_3),$$

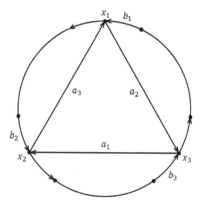

FIGURE 3.6: Digraph D.

where

$$\mathcal{E}^*(D) = \mathcal{E}(D, \bar{a}_1, a_2, a_3) \dot{\cup} \mathcal{E}(D, \bar{a}_2, a_1, a_3) \dot{\cup} \mathcal{E}(D, \bar{a}_3, a_1, a_2)$$
$$\dot{\cup} \mathcal{E}(D, a_1, \bar{a}_2, \bar{a}_3) \dot{\cup} \mathcal{E}(D, a_2, \bar{a}_1, \bar{a}_3) \dot{\cup} \mathcal{E}(D, a_3, \bar{a}_1, \bar{a}_2).$$

Let $D' = D - A(T)$, where $A(T)$ stands for the arc set of T. Then D' is an Eulerian digraph whose arc set is decomposed into a Hamiltonian cycle and $n-1$ pairwise vertex disjoint triangles. By induction hypothesis, $|\mathcal{E}(D')| \equiv 2 \pmod 4$. It is obvious that

$$|\mathcal{E}(D')| = |\mathcal{E}(D, a_1, a_2, a_3)| = |\mathcal{E}(D, \bar{a}_1, \bar{a}_2, \bar{a}_3)|.$$

So $|\mathcal{E}(D, a_1, a_2, a_3)| + |\mathcal{E}(D, \bar{a}_1, \bar{a}_2, \bar{a}_3)| \equiv 0 \pmod 4$. As

$$|\mathcal{E}(D)| = |\mathcal{E}^*(D)| + |\mathcal{E}(D, a_1, a_2, a_3)| + |\mathcal{E}(D, \bar{a}_1, \bar{a}_2, \bar{a}_3)|,$$

we have $|\mathcal{E}(D)| \equiv |\mathcal{E}^*(D)| \pmod 4$. So it remains to show that $|\mathcal{E}^*(D)| \equiv 2 \pmod 4$.

For $i = 1, 2, 3$, let x_i^- be the predecessor of x_i on C and let $b_i = (x_i^-, x_i)$. By Corollary 3.1.13, we may assume that the cycle C is directed as in Figure 3.6. For $i = 1, 2, 3$, let C_i be the unique directed cycle in $C \cup a_i$ containing a_i.

For $(i, j, k) \in \{(1, 2, 3), (2, 3, 1), (3, 1, 2)\}$, let $D_i = (D - C_i) \cup C_i^R$ as in Figure 3.7 and let D_i' be obtained from D_i by splitting away arcs b_j, a_k, a_j and b_k^R as in Figure 3.8. Let x_i', x_j', x_k' be the new vertices that result from the splitting. Then D_i' is an Eulerian digraph which has a decomposition into a directed Hamiltonian cycle and $(n-1)$ pairwise vertex disjoint triangles.

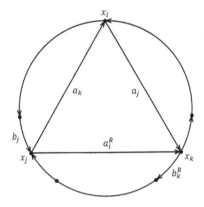

FIGURE 3.7: Digraph D_i.

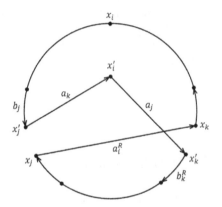

FIGURE 3.8: Digraph D_i'.

Since a_j and a_k share a degree 2 vertex, each Eulerian subdigraph H of D_i' either contains both a_j and a_k, or contains neither of a_j and a_k. Hence

$$
\begin{aligned}
|\mathcal{E}(D_i')| \;=\; & |\mathcal{E}(D_i', a_i^R, a_j, a_k)| + |\mathcal{E}(D_i', \overline{a}_i^R, a_j, a_k)| \\
& + |\mathcal{E}(D_i', a_i^R, \overline{a}_j, \overline{a}_k)| + |\mathcal{E}(D_i', \overline{a}_i^R, \overline{a}_j, \overline{a}_k)|.
\end{aligned}
$$

Claim 3.4.3 *For* $(i,j,k) \in \{(1,2,3),(2,3,1),(3,1,2)\}$,

$$
\begin{aligned}
|\mathcal{E}(D_i', a_i^R, a_j, a_k)| &= |\mathcal{E}(D, \overline{a}_i, a_j, a_k)|, \\
|\mathcal{E}(D_i', \overline{a}_i^R, \overline{a}_j, \overline{a}_k)| &= |\mathcal{E}(D, a_i, \overline{a}_j, \overline{a}_k)|, \\
|\mathcal{E}(D_i', a_i^R, \overline{a}_j, \overline{a}_k)| &= |\mathcal{E}(D', \overline{b}_j, b_k)| = |\mathcal{E}(D', b_i, \overline{b}_j, b_k)| \\
&\quad + |\mathcal{E}(D', \overline{b}_i, \overline{b}_j, b_k)|, \\
|\mathcal{E}(D_i', \overline{a}_i^R, a_j, a_k)| &= |\mathcal{E}(D', b_j, \overline{b}_k)| = |\mathcal{E}(D', b_i, b_j, \overline{b}_k)| \\
&\quad + |\mathcal{E}(D', \overline{b}_i, b_j, \overline{b}_k)|.
\end{aligned}
$$

Proof. For $H \in \mathcal{E}(D)$, let $\phi(H) = (H - C_i) \cup (C_i - H)^R$. Then ϕ is a one-to-one correspondence between $\mathcal{E}(D, \overline{a}_i, a_j, a_k)$ and $\mathcal{E}(D_i', a_i^R, a_j, a_k)$, and a one-to-one correspondence between $\mathcal{E}(D, a_i, \overline{a}_j, \overline{a}_k)$ and $\mathcal{E}(D_i', \overline{a}_i^R, \overline{a}_j, \overline{a}_k)$.

For $H \in \mathcal{E}(D_i')$, $\psi(H) = (H - C_i^R) \cup (C_i^R - H)^R$ is a one-to-one correspondence between $\mathcal{E}(D_i', a_i^R, \overline{a}_j, \overline{a}_k)$ and $\mathcal{E}(D', \overline{b}_j, b_k)$, and $\psi'(H) = \psi(H) - \{a_i, a_j, a_k\}$ is a one-to-one correspondence between $\mathcal{E}(D_i', \overline{a}_i^R, a_j, a_k)$ and $\mathcal{E}(D', b_j, \overline{b}_k)$. ∎

It follows from Claim 3.4.3 that

$$
|\mathcal{E}(D_1')| + |\mathcal{E}(D_2')| + |\mathcal{E}(D_3')| = |\mathcal{E}^*(D)| + m,
$$

where

$$
\begin{aligned}
m = \sum_{(i,j,k)} \big(&|\mathcal{E}(D', b_i, \overline{b}_j, b_k)| + |\mathcal{E}(D', \overline{b}_i, \overline{b}_j, b_k)| \\
&+ |\mathcal{E}(D', b_i, b_j, \overline{b}_k)| + |\mathcal{E}(D', \overline{b}_i, b_j, \overline{b}_k)| \big),
\end{aligned}
$$

where the summation is over $(i,j,k) \in \{(1,2,3),(2,3,1),(3,1,2)\}$.

By induction hypothesis, $|\mathcal{E}(D_i')| \equiv 2 \pmod 4$ and hence

$$
|\mathcal{E}(D_1')| + |\mathcal{E}(D_2')| + |\mathcal{E}(D_3')| \equiv 2 \pmod 4.
$$

To prove that $|\mathcal{E}^*(D)| \equiv 2 \pmod 4$, it remains to show that $m \equiv 0 \pmod 4$.

By using the one-to-one correspondence $\phi : \mathcal{E}(D') \to \mathcal{E}(D')$ defined as $\phi(H) = D \backslash H$, we conclude that $|\mathcal{E}(D', b_i, \overline{b}_j, \overline{b}_k)| = |\mathcal{E}(D', \overline{b}_i, b_j, b_k)|$. It follows that

$$
\begin{aligned}
&|\mathcal{E}(D', b_i, \overline{b}_j, b_k)| + |\mathcal{E}(D', \overline{b}_i, \overline{b}_j, b_k)| + |\mathcal{E}(D', b_i, b_j, \overline{b}_k)| \\
&+ |\mathcal{E}(D', \overline{b}_i, b_j, \overline{b}_k)| = 2(|\mathcal{E}(D', b_i, \overline{b}_j, b_k)| + |\mathcal{E}(D', b_i, b_j, \overline{b}_k)|).
\end{aligned}
$$

Therefore

$$m = 4(|\mathcal{E}(D',\bar{b}_1,b_2,b_3)| + |\mathcal{E}(D',b_1,\bar{b}_2,b_3)| + |\mathcal{E}(D',b_1,b_2,\bar{b}_3)|)$$
$$\equiv 0 \pmod 4.$$

This completes the proof of Lemma 3.4.2, and hence completes the proof of Theorem 3.4.1. ∎

Theorem 3.4.1 was extended in [24], where it was proved that if $k \geq 4$, and G is a graph whose edge set can be decomposed into a Hamilton cycle and a family of vertex disjoint complete graphs of order at most k, then G is k-colourable. A different proof of Theorem 3.3.2 is given in [49].

3.5 Bound for $AT(G)$ in terms of degree

It is obvious that for any graph G, $\mathrm{col}(G) \leq \Delta(G)+1$. Moreover if G is connected, and G is not a regular graph, we have $\mathrm{col}(G) \leq \Delta(G)$. However if G is a regular graph, then $\mathrm{col}(G) = \Delta(G) + 1$. Theorem 3.5.3 below implies that even if G is a regular graph, we still have $AT(G) \leq \Delta(G)$, provided that G is neither a complete graph nor an odd cycle.

A graph G is called *degree choosable* if for any list assignment L for which $|L(v)| = d_G(v)$ for every vertex v, then G is L-colourable. We say that G is *degree AT* if G is f-AT for $f(v) = d_G(v)$ for every vertex v of G. By CNS, if a graph G is degree-AT, then it is degree choosable. A connected graph G is called a *Gallai tree* if every block of G is either a complete graph or an odd cycle.

FIGURE 3.9: A Gallai tree.

Lemma 3.5.1 *If G is a Gallai tree, then G is not degree choosable, and hence not degree-AT.*

Proof. Assume that G is a Gallai tree. The proof is by induction on the number of blocks of G. If G is a complete graph K_n, then each vertex has degree $n-1$, and G is not $(n-1)$-colourable, and hence not $(n-1)$-choosable. If G is an odd cycle, then each vertex has degree 2 and G is not 2-colourable, and hence not 2-choosable.

Assume that G has more than one block. Let B be a leaf block of G, i.e., B contains exactly one cut-vertex v. Let $G' = G - (V(B) - \{v\})$. Then G' is a Gallai tree and hence is not degree choosable. Let L' be a list assignment of G' with $|L'(x)| = d_{G'}(x)$ for each vertex x of G' so that G' is not L'-colourable. Assume that B is k-regular and let C be a set of k colours disjoint from $L'(v)$. Let $L(x) = L'(x)$ for $x \in V(G') - \{v\}$, $L(v) = L'(v) \cup C$ and $L(x) = C$ for $x \in V(B) - \{v\}$. Then $|L(x)| = d_G(x)$ for every vertex x.

It remains to show that G is not L-colourable. Assume to the contrary that ϕ is an L-colouring of G. If $\phi(v) \in C$, then the restriction of ϕ to B is a k-colouring of B, a contradiction. If $\phi(v) \notin C$, then the restriction of ϕ to G' is an L'-colouring of G', again a contradiction. ∎

It was proved in [28] that the converse of Lemma 3.5.1 is also true. Before proving this result, we first prove the following lemma.

Lemma 3.5.2 *If G is a 2-connected graph which is neither complete nor an odd cycle, then G contains an even cycle which contains at most one chord.*

Proof. Let S be a minimum vertex cut of G. As G is 2-connected, $|S| \geq 2$. Let u, v be two vertices of S. Let G_1, G_2 be the two induced subgraphs of G with $V(G_1) \cap V(G_2) = S$ and $V(G_1) \cup V(G_2) = V(G)$. Let P_i be a shortest path in $G_i - (S - \{u, v\}) - E(S)$ connecting u and v. Then each P_i has length at least 2, and the union $C = P_1 \cup P_2$ is a cycle of length at least 4. If C is an even cycle, then we are done, as C has at most one chord uv. So assume that C is an odd cycle. If $e = uv$ is an edge, then either $P_1 \cup e$ or $P_2 \cup e$ is an induced even cycle and we are done. Assume therefore that uv is not an edge of G. So C is an induced odd cycle. As G is not an odd cycle, we may assume that $G_1 - C \neq \emptyset$. If each vertex of $G_1 - C$ has at most one neighbour in C, then let P_3 be a shortest path in $G_1 - E(C)$ connecting two vertices of C. Then the union $P_1 \cup P_2 \cup P_3$ consists of three paths joining the two end vertices of P_3. Hence it contains an induced even cycle, and we are done. Thus we may assume that $G_1 - C$ has a vertex

w which has at least two neighbours in C. Let $u_1, u_2, \ldots, u_k$ be the neighbours of w in C, in this order. For $i = 1, 2, \ldots, k$, let Q_i be the subpath of C connecting u_i and u_{i+1} (with $u_{k+1} = u_1$). If Q_i has even length, then $Q_i \cup \{w\}$ induces an even cycle and we are done. So assume that each Q_i has odd length. If $k = 3$, $|Q_1| = 3$, $|Q_2| = |Q_3| = 1$, then $Q_2 \cup C_3 \cup w$ is an even cycle with one chord. Otherwise, $Q_1 \cup Q_2 \cup \{w\}$ induces an even cycle with one chord wu_2. ∎

Theorem 3.5.3 *A connected graph G is degree-AT if and only if G is not a Gallai tree.*

Proof. By Lemma 3.5.1, it suffices to show that if G is not a Gallai tree, then G is degree-AT.

Assume that G is a connected graph which is not a Gallai tree. By Lemma 3.5.2, G has an even cycle C which contains at most one chord.

Let G' be obtained from G by contracting C into a single vertex w. Order the vertices of G' so that x precedes y if $d_{G'}(y, w) > d_{G'}(x, w)$, and order x, y arbitrarily if $d_{G'}(y, w) = d_{G'}(x, w)$. Then orient edge $e = xy$ from x to y if x precedes y. Observe that each vertex other than w has at least one in-neighbour. Extend this orientation to G, so that C is oriented as a directed cycle, and if C has a chord, then the chord is oriented arbitrarily. Denote the resulting orientation by D. Then each vertex has at least one in-neighbour. So $d_D^+(x) < d_G(x)$ for each vertex x. There are at least two even Eulerian subdigraphs in D: the empty subdigraph and the directed cycle C. On the other hand, there is at most one odd Eulerian subgraph in D, which is induced by the chord of C together with one path of C connecting the two ends of the chord. So $|\mathcal{E}_e(D)| - |\mathcal{E}_o(D)| \neq 0$. As $|L(x)| = d_G(x) > d_D^+(x)$, we conclude that G is degree-AT. ∎

3.6 Planar graphs

It was proved first by Voigt [62] that there are planar graphs that are not 4-choosable. The following example is modified from an example given by Choi and Kwon [17].

Theorem 3.6.1 *There exists a non-4-choosable planar graph.*

Proof. Let H be the graph, and L, the list assignment of H as given in Figure 3.10.

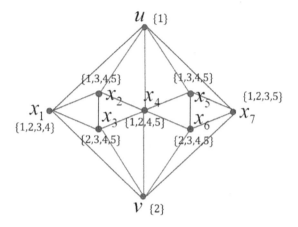

FIGURE 3.10: The graph H.

First we show that H is not L-colourable. Assume to the contrary that ϕ is an L-colouring of H. Since $\phi(u) = 1$ and $\phi(v) = 2$, we conclude that $\phi(x_4) \in \{4,5\}$. If $\phi(x_4) = 5$, then each of the three vertices in the triangle $\{x_1, x_2, x_3\}$ needs to be coloured by colours 3 and 4, a contradiction. If $\phi(x_4) = 4$, then each of the three vertices in the triangle $\{x_5, x_6, x_7\}$ needs to be coloured by colours 3 and 5, again a contradiction.

Now let G be the planar graph obtained from K_4 by adding, for every ordered pair of vertices (x, y) of K_4, a copy of H, identifying u with x and identifying v with y. Let $L(x) = \{1, 2, 3, 4\}$ for each vertex x of K_4, and for each copy of H and for each $x \in V(H) - \{u, v\}$, let $L(x)$ be as defined above. Then L is a 4-list assignment of G. Assume that ϕ is an L-colouring of G. Then one ordered pair of vertices of K_4 is coloured by 1 and 2. But such a colouring cannot be extended to an L-colouring of the copy of H associated with this pair of vertices. So G is not L-colourable. ∎

This non-4-choosable planar graph has 88 vertices. The smallest known non-4-choosable planar graph, due to Mirzakhani [48], has 63 vertices.

Since G is not 4-choosable, we have $AT(G) \geq 5$. The following theorem shows that $AT(G) \leq 5$ for every planar graph G.

Theorem 3.6.2 *[67] If G is a planar graph, then $AT(G) \leq 5$.*

Recall that $ch(G) \leq AT(G)$ so that Theorem 3.6.2 implies the following well-known result of Carsten Thomassen [60]: For every planar graph G, $ch(G) \leq 5$. In other words, every planar graph is 5-choosable.

Theorem 3.6.2 is a consequence of the following more technical result.

Theorem 3.6.3 *Assume that G is a plane graph and $e = v_1 v_2$ is a boundary edge of G. Let $f(v_1) = f(v_2) = 1$, $f(v) = 3$ for every other boundary vertex v and $f(v) = 5$ for every interior vertex v. Then $G - e$ is f-AT.*

Proof. As per our assumption, G is a plane graph and $e = v_1 v_2$ is a boundary edge of G. Let $f : V(G) \to Z$ be defined as above. An AT-orientation D of $G - e$ with $d_D^+(v) \leq f(v) - 1$ is called a *nice orientation* for (G, e). We shall prove that for any planar graph G and for any boundary edge e of G, (G, e) has a nice orientation.

Assume that the theorem is not true and G is a minimum counterexample. If G has a cut vertex v, then let G_1, G_2 be induced subgraphs of G with $V(G_1) \cap V(G_2) = \{v\}$ and $V(G_1) \cup V(G_2) = V(G)$. Assume that $e \in E(G_1)$. Let u be a boundary neighbour of v in G_2 and $e' = vu$. By the minimality of G, (G_1, e) has a nice orientation D_1, and (G_2, e') has a nice orientation D_2. Let D be obtained from $D_1 \cup D_2$ by adding the arc (u, v). Edges in $D_2 \cup \{(u, v)\}$ incident to v are not contained in any directed cycles (since $d_{D_2}^+(v) = 0$), and hence are not contained in any Eulerian subdigraph of D. Therefore

$$|\mathcal{E}_e(D)| = |\mathcal{E}_e(D_1)| \times |\mathcal{E}_e(D_2)| + |\mathcal{E}_o(D_1)| \times |\mathcal{E}_o(D_2)|,$$
$$|\mathcal{E}_o(D)| = |\mathcal{E}_e(D_1)| \times |\mathcal{E}_o(D_2)| + |\mathcal{E}_o(D_1)| \times |\mathcal{E}_e(D_2)|.$$

So $\text{diff}(D) = \text{diff}(D_1) \times \text{diff}(D_2) \neq 0$. Hence D is a nice orientation of (G, e).

Thus we may assume that G is 2-connected and the boundary of G is a simple cycle $B_G = (v_1, v_2, \ldots, v_n)$.

If B_G has a chord $e' = xy$, then let G_1, G_2 be the two e'-components with $e \in G_1$ (i.e., G_1, G_2 are induced subgraphs with $V(G_1) \cap V(G_2) = \{x, y\}$ and $V(G_1) \cup V(G_2) = V(G)$). By the induction hypothesis, (G_1, e) has a nice orientation D_1, and (G_2, e') has a nice orientation D_2. Let $D = D_1 \cup D_2$. The same argument as above shows that D is a nice orientation of (G, e).

Assume that G has no chord. Let $G' = G - v_n$. Let $v_1, u_1, u_2, \ldots, u_k, v_{n-1}$ be the neighbours of v_n. By induction hypothesis, (G', e) has a nice orientation D'. Let D be obtained from D' by adding the arcs (v_n, v_1) and (v_n, v_{n-1}) and (u_i, v_n) for $i = 1, 2, \ldots, k$. We denote the arc (u_i, v_n) by a_i. Then D satisfies the out-degree condition for being a nice orientation of (G, e). So if D is an AT-orientation, then we are done.

Assume that D is not an AT-orientation. For $i = 1, 2, \ldots, k$, let $\mathcal{E}(D, a_i)$ denote the set of Eulerian subdigraphs H of D which contains the arc $a_i = (u_i, v_n)$. Then

$$\mathcal{E}_e(D) = \mathcal{E}_e(D') \cup \bigcup_{i=1}^k \mathcal{E}_e(D, a_i), \text{ and } \mathcal{E}_o(D) = \mathcal{E}_o(D') \cup \bigcup_{i=1}^k \mathcal{E}_o(D, a_i).$$

Since v_n has only one out arc, namely (v_n, v_{n-1}), which can be contained in a directed cycle, each Eulerian sub-digraph contains at most one in arc of v_n. Thus for $i \neq j$,

$$\mathcal{E}_e(D, a_i) \cap \mathcal{E}_e(D, a_j) = \emptyset, \ \mathcal{E}_o(D, a_i) \cap \mathcal{E}_o(D, a_j) = \emptyset,$$

i.e., the unions above are disjoint unions. So

$$
\begin{aligned}
|\mathcal{E}_e(D)| &= |\mathcal{E}_e(D')| + \sum_{i=1}^k |\mathcal{E}_e(D, a_i)|, \text{ and } |\mathcal{E}_o(D)| = |\mathcal{E}_o(D')| \\
&+ \sum_{i=1}^k |\mathcal{E}_o(D, a_i)|.
\end{aligned}
$$

Hence

$$0 = \operatorname{diff}(D) = \operatorname{diff}(D') + \sum_{i=1}^k \operatorname{diff}(\mathcal{E}(D, a_i)).$$

As $\operatorname{diff}(D') \neq 0$, there exists $i \in \{1, 2, \ldots, k\}$ such that $\operatorname{diff}(\mathcal{E}(D, a_i)) \neq 0$.

Assume that $\operatorname{diff}(\mathcal{E}(D, a_i)) \neq 0$, where $i \in \{1, 2, \ldots, k\}$. Then D has a directed cycle C containing a_i. Let D'' be obtained from $(D-C) \cup C^R$ by reversing the arc a_i^R (thus the edge $u_i v_n$ is oriented as (u_i, v_n) in D''). It is straightforward to verify that D'' satisfies the out-degree conditions for being a nice orientation for (G, e). However, we shall show that D'' is an AT-orientation, and hence D'' is indeed a nice orientation of (G, e).

Observe that no arc incident to v_n is contained in a directed cycle of D''. Hence none of these arcs is contained in an Eulerian subdigraph of D''. For each Eulerian subdigraph H of D'', let $\phi(H) = (H - C^R) \cup (C^R - H)^R$. Then $\phi(H)$ is an Eulerian subdigraph of D containing a_i. Conversely, if H is an Eulerian subdigraph of D containing arc a_i, then $\psi(H) = (H - C) \cup (C - H)^R$ is an Eulerian subdigraph of D''. Further, $\psi(\phi(H)) = H$. So ϕ is a one-to-one correspondence between $\mathcal{E}(D'')$ and $\mathcal{E}(D, a_i)$. Moreover, if C is an even cycle, then $\phi(H)$ and H have the same parity, and if C is an odd cycle, then $\phi(H)$ and H have opposite parities. Therefore $\mathrm{diff}(D'') = \pm\mathrm{diff}(\mathcal{E}(D, a_i)) \neq 0$. Hence D'' is an AT-orientation and hence a nice orientation of (G, e).

This completes the proof of Theorem 3.6.3. ∎

3.7 Planar graph minus a matching

A *d-defective colouring* of a graph G is a colouring of the vertices of G such that each colour class induces a subgraph of maximum degree at most d. Thus, a 0-defective colouring of G is simply a proper colouring of G, while in a 1-defective colouring, a matching is allowed as a set of non-properly coloured edges. We say G is *d-defective k-colourable* if there is a d-defective colouring of G using a total number of k colours.

Given a k-list assignment L of G, a *d-defective L-colouring* of G is a d-defective colouring c of G with $c(v) \in L(v)$ for every vertex v of G. A graph G is *d-defective k-choosable* if for any k-list assignment L of G, there exists a d-defective L-colouring of G. Clearly, every d-defective k-choosable graph is d-defective k-colourable. The converse is not true.

It was proved by Cowen, Cowen and Woodall in [18] that every planar graph is 2-defective 3-colourable, and every outerplanar graph is 2-defective 2-colourable. Eaton and Hull [21], and Škrekovski [58] independently extended these results to list colouring, and showed that every planar graph is 2-defective 3-choosable, and every outerplanar graph is 2-defective 2-choosable. Cushing and Kierstead [19] proved that every planar graph is 1-defective 4-choosable. The result of Eaton and Hull and Škrekovski, and the result of Cushing and Kierstead can be formulated as follows:

Assume that G is a planar graph. For any 3-list assignment L of G, there is a subgraph H of G with $\Delta(H) \leq 2$ such that $G - E(H)$ is L-colourable; For any 4-list assignment L of G, there is a subgraph H of G with $\Delta(H) \leq 1$ such that $G - E(H)$ is L-colourable.

The subgraph H of G depends on the list assignment L. It is natural to ask whether one can choose a subgraph H of G independent of the list assignment L. In other words, we have the following questions:

Question 3.7.1 *Does every planar graph G have a subgraph H with $\Delta(H) \leq 2$ such that $G \backslash E(H)$ is 3-choosable?*

Question 3.7.2 *Does every planar graph G have a subgraph H with $\Delta(H) \leq 1$ such that $G \backslash E(H)$ is 4-choosable?*

It turns out that the answer to Question 3.7.1 is negative, and the answer to Question 3.7.2 is positive. First we show that there is a planar graph G such that for any subgraph H of G with $\Delta(H) \leq 3$, $G \backslash E(H)$ is not 3-choosable. The following example was given in [35].

Let J_1 and J_2 be the two graphs depicted in Figure 3.11. For $i = 1, 2$, the edge ab in J_i is called the *handle* of J_i.

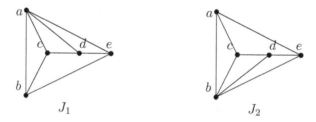

FIGURE 3.11: The graphs J_1 and J_2.

Let $\mathcal{J}$ be the set of graphs obtained from the disjoint union of 6 copies of J_1 or J_2 by identifying the edges corresponding to ab from each copy. For each $G \in \mathcal{J}$, let c_i, d_i, e_i be the vertices corresponding to c, d, e, respectively, for $i \in [6]$, and the edge ab in G is called the *handle* of G. (See Figure 3.12.)

Lemma 3.7.3 *For $i = 1, 2$, there is a list assignment L of J_i such that $|L(a)| = |L(b)| = 1$, $|L(c)| = |L(d)| = |L(e)| = 3$ and J_i is not L-colourable.*

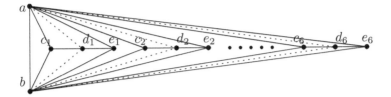

FIGURE 3.12: This illustrates a graph in $\mathcal{J}$. For each $i \in [6]$, the subgraph induced on $\{a, b, c_i, d_i, e_i\}$ is isomorphic to J_1 or J_2 with isomorphism mapping a, b, c_i, d_i, e_i to a, b, c, d, e, respectively. That is, there exists exactly one edge between $\{a, b\}$ and d_i.

Proof. We consider the graph J_1. Let $L(a) = \{1\}, L(b) = \{2\}, L(c) = \{1, 2, 3\}, L(d) = \{1, 3, 4\}, L(e) = \{1, 2, 4\}$. It is straightforward to check that J_1 is not L-colourable. ∎

Lemma 3.7.4 *None of the graphs* $G \in \mathcal{J}$ *is 3-choosable.*

Proof. We construct a 3-list assignment of G as follows: $L(a) = L(b) = \{\alpha, \beta, \gamma\}$. For each pair (x, y) of distinct colours from $\{\alpha, \beta, \gamma\}$, there is a copy of J_1 or J_2 whose lists are given in such a way that there is no L-colouring ϕ of this copy of J_1 or J_2 with $\phi(a) = x$ and $\phi(b) = y$. By Lemma 3.7.3, such a list assignment exists. Therefore G is not L-colourable.

To be precise, the 3-list assignment L of G is defined as follows. Let $(x_i, y_i, z_i)_{i=1,\ldots,6}$ be the six permutations of the colour set $\{\alpha, \beta, \gamma\}$.

- $L(a) = L(b) = \{\alpha, \beta, \gamma\}$.

- For each $i \in [6]$, $L(c_i) = \{\alpha, \beta, \gamma\}$ and $L(e_i) = \{x_i, y_i, \omega\}$.

- For each $i \in [6]$, $L(d_i) = \{x_i, z_i, \omega\}$ if d_i is adjacent to a and $L(d_i) = \{y_i, z_i, \omega\}$ if d_i is adjacent to b.

Suppose there exists an L-colouring ϕ of G. We may assume that $\phi(a) = \alpha$ and $\phi(b) = \beta$. Without loss of generality, let $x_1 = \alpha$ and $y_1 = \beta$. Since $L(c_1) = \{\alpha, \beta, \gamma\}$ and $L(e_1) = \{x_1, y_1, \omega\} = \{\alpha, \beta, \omega\}$, we have $\phi(c_1) = \gamma$ and $\phi(e_1) = \omega$. If d_1 is adjacent to a, then $L(d_1) = \{\alpha, \gamma, \omega\}$ but $\phi(a) = \alpha$, $\phi(c_1) = \gamma$ and $\phi(e_1) = d$, so there is no available colour for d_1. Similarly, if d_1 is adjacent to b, then $L(d_1) = \{\beta, \gamma, \omega\}$ but $\phi(b) = \beta$, $\phi(c_1) = \gamma$ and $\phi(e_1) = \omega$, so there is no available colour for d_1. By the construction of

G, d_1 is adjacent to either a or b, therefore, there is no possible colour for d_1. This leads to a contradiction. Therefore G is not 3-choosable. ∎

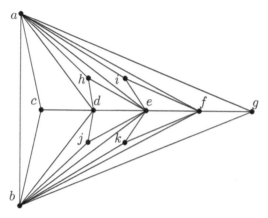

FIGURE 3.13: The graph J_3.

Lemma 3.7.5 *Assume that H is a subgraph of J_3 (the graph depicted in Figure 3.13) with maximum degree at most three. If H does not contain any edge incident with a or b, then $J_3 - E(H)$ contains a copy of K_4, or a subgraph isomorphic to J_1 or J_2 with handle ab.*

Proof. Assume that H is a subgraph of J_3 which does not contain any edge incident with a or b. If any of the edges cd, de, ef, fg is not contained in H, then $J_3 - E(H)$ contains a copy of K_4 and we are done. Thus we assume that H contains $\{cd, de, ef, fg\}$. If H contains neither hd nor he, then the edge set $\{ab, ad, ah, ae, hd, he, bd, be\}$ induces a copy of J_1 with handle ab in $J_3 - E(H)$, and we are done. So we assume that H contains hd or he. Similarly, assume that H contains ie or if, jd or je, and ke or kf. Therefore $|E(H) \cap \{hd, he, ie, if, jd, je, ke, kf\}| \geq 4$. Since every edge in $\{hd, he, ie, if, jd, je, ke, kf\}$ is incident with d, e or f, it follows from the pigeonhole principle that one of d, e and f is an end of at least two edges in $E(H) \cap \{hd, he, ie, if, jd, je, ke, kf\}$. It implies that one of d, e and f has degree at least four in H because the edges cd, de, ef and fg are already contained in H, which yields a contradiction. This completes the proof. ∎

Let S be the graph obtained from nine copies of J_3 by identifying the edges corresponding to ab from each copy. It is obvious that S is a planar graph. The edge ab in S is called the *handle* of S. We obtain the following as a corollary of Lemma 3.7.5.

Corollary 3.7.6 *Assume that H is a subgraph of S with maximum degree at most three. If H does not contain any edge incident with a in S, then $S - E(H)$ contains a copy of K_4 or a member of $\mathcal{J}$ as a subgraph.*

Proof. Let $S_1, S_2, \ldots, S_9$ be the distinct subgraphs of S isomorphic to J_3 with handle ab. Since H has maximum degree at most three, and H does not contain any edge incident with a in S, there are $1 \le i_1 < \ldots < i_6 \le 9$ such that every edge incident with b in S_{i_j} is not contained in H for $j \in [6]$. Without loss of generality, let $i_j = j$ for $j \in [6]$. Then, by Lemma 3.7.5, for each $j \in [6]$, $S_j - E(H)$ contains a copy of K_4 or a subgraph isomorphic to J_1 or J_2 with handle ab. We are done if $S_j - E(H)$ contains K_4, so, we may assume that $S_j - E(H)$ has a subgraph S'_j isomorphic to J_1 or J_2 with handle ab. Then, combining S'_j for $j \in [6]$, we obtain a subgraph of $S - E(H)$ isomorphic to a member of $\mathcal{J}$. This completes the proof. ∎

Now, we construct a planar graph G such that for every subgraph H of G with maximum degree at most 3, $G - E(H)$ is not 3-choosable: Start with a star with four leaves v_1, v_2, v_3, v_4 and centre c, and for each $i \in [4]$, we add a copy of S with handle cv_i to the star. That is, G consists of four edge-disjoint copies S_1, S_2, S_3, S_4 of S where the handle of S_i is cv_i for $i \in [4]$.

Theorem 3.7.7 *For every subgraph H in G_1 with $\Delta(H) \le 3$, $G_1 - E(H)$ is not 3-choosable.*

Proof. Let H be a subgraph in G with $\Delta(H) \le 3$. We claim that $G - E(H)$ contains K_4 or a member of $\mathcal{J}$. Then, by the fact that K_4 is not 3-colourable (so not 3-choosable) and Lemma 3.7.4, Theorem 3.7.7 follows.

By adding isolated vertices to H, we consider H as a spanning subgraph of G with maximum degree at most three. Since c has degree at most three in H, there exists $i \in [4]$ such that H does not contain any edge incident with c in S_i. Then, by Corollary 3.7.6, $S_i - E(H)$ contains K_4 or a member of $\mathcal{J}$. This completes the proof. ∎

Next we show that Question 3.7.2 has a positive answer.

Theorem 3.7.8 *Every planar graph* G *contains a matching* M *such that* $AT(G \backslash M) \le 4$.

Theorem 3.7.8 was proved in [26], where the proof is a direct calculation of the coefficient of a certain monomial in the expansion of $f_G(\mathbf{x})$. Here we present a proof which is basically the same as the proof of Theorem 3.6.2, except that the technical statement is different.

Definition 3.7.9 *Assume that* G *is a plane graph,* $e = v_1 v_2$ *is a boundary edge of* G, *and* M *is a matching in* G *containing* e. *An orientation* D *of* $G \backslash M$ *is* nice *for* (G, e, M) *if the following conditions hold:*

1. *D is an AT-orientation.*

2. *$d_D^+(v_1) = d_D^+(v_2) = 0$.*

3. *$d_D^+(v) \le 2 - d_M(v)$ for every other boundary vertex v.*

4. *$d_D^+(v) \le 3$ for each interior vertex v.*

Notice that $d_M(v) = 1$ if v is covered by M, and $d_M(v) = 0$ otherwise.

Theorem 3.7.8 follows from Theorem 3.7.10 given below.

Theorem 3.7.10 *Assume that* G *is a plane graph and* $e = v_1 v_2$ *is a boundary edge of* G. *Then* G *has a matching* M *containing* e *such that there exists a nice orientation* D *for* (G, e, M).

Proof. Assume that the theorem is false and G is a minimum counterexample. The same argument as in the proof of Theorem 3.6.3 shows that G has no cut-vertex and the boundary cycle B_G of G has no chord.

Let $G' = G - v_n$. Let $v_1, u_1, u_2, \ldots, u_k, v_{n-1}$ be the neighbours of v_n. By induction hypothesis, G' has a matching M' such that (G', e, M') has a nice orientation D'. Let D be obtained from D' by adding arcs (v_n, v_1) and (v_n, v_{n-1}) and arcs $a_i = (u_i, v_n)$ for $i = 1, 2, \ldots, k$. If D is an AT-orientation, then it is a nice orientation for (G, e, M') and we are done.

Assume that D is not an AT-orientation. For $i = 1, 2, \ldots, k$, $\mathcal{E}(D, a_i)$ is the set of Eulerian subdigraphs H of D which contains the arc $a_i = (u_i, v_n)$. Then

$$\mathcal{E}_e(D) = \mathcal{E}_e(D') \cup \bigcup_{i=1}^{k} \mathcal{E}_e(D, a_i), \text{ and } \mathcal{E}_o(D) = \mathcal{E}_o(D') \cup \bigcup_{i=1}^{k} \mathcal{E}_o(D, a_i).$$

Moreover the unions are disjoint unions. So

$$|\mathcal{E}_e(D)| = |\mathcal{E}_e(D')| + \sum_{i=1}^{k} |\mathcal{E}_e(D, a_i)|, \text{ and } |\mathcal{E}_o(D)| = |\mathcal{E}_o(D')|$$

$$+ \sum_{i=1}^{k} |\mathcal{E}_o(D, a_i)|.$$

Now $|\mathcal{E}_e(D)| = |\mathcal{E}_o(D)|$ and $|\mathcal{E}_e(D')| \neq |\mathcal{E}_o(D')|$ imply that $|\mathcal{E}_e(D, a_i)| \neq |\mathcal{E}_o(D, a_i)|$ for some $i \in \{1, 2, \ldots, k\}$.

Assume that $|\mathcal{E}_e(D, a_i)| \neq |\mathcal{E}_o(D, a_i)|$, where $i \in \{1, 2, \ldots, k\}$. Then D has a directed cycle C containing a_i. If $d_D^+(u_i) \leq 1$, then let D'' be obtained from $(D - C) \cup C^T$ by reversing the arc a_i^R (thus the edge $u_i v_n$ is oriented as (u_i, v_n) in D''). The same argument as in the proof of Theorem 3.6.2 shows that D'' is an AT-orientation. As $d_{D''}^+(u_i) = d_D^+(u_i) + 2 \leq 3$, D'' satisfies the out-degree conditions of being a nice orientation for (G, e, M'), and hence D'' is indeed a nice orientation for (G, e, M'). Otherwise $d_D^+(u_i) = 2$. As $d_D^+(u_i) \leq 2 - d_{M'}(u_i)$, we conclude that $d_{M'}(u_i) = 0$, i.e., u_i is not covered by M'. Let $M = M' \cup \{u_i v_n\}$. Then M is a matching in G containing e. Let D'' be obtained from $(D - C) \cup C^T$ by deleting the arc a_i^R. Then $d_{D''}^+(u_i) = d_D^+(u_i) + 1 = 3$, and hence D'' satisfies the out-degree conditions of being a nice orientation for (G, e, M). Again, the same argument as in the proof of Theorem 3.6.2 shows that D'' is an AT-orientation. Hence D'' is indeed a nice orientation for (G, e, M). ∎

3.8 Discharging method

The following lemma is a common tool used in proving that a graph G has an f-AT orientation D for some $f : V(G) \to \mathbb{N}$.

Lemma 3.8.1 *Assume that G is a graph, X is a subset of $V(G)$ and $f : V(G) \to \mathbb{N}$ is a mapping which assigns to each vertex v a positive integer $f(v)$. For $v \in X$, let $s(v)$ be the number of edges of G connecting v and $V(G) \backslash X$ and let $g(v) = f(v) - s(v)$. If $G - X$ has an f-AT orientation, and $G[X]$ has a g-AT orientation, then G has an f-AT orientation.*

Proof. Let D_1 be an f-AT orientation of $G-X$ and D_2 be a g-AT-orientation of $G[X]$. Let D be obtained from $D_1 \cup D_2$ by orienting all the edges between X and $V(G)\backslash X$ from X to $V(G)\backslash X$. Then for each vertex v of G, we have $d_D^+(v) \le f(v)-1$. Since no arcs between X and $V(G)\backslash X$ is contained in a directed cycle of D, and hence not contained in any Eulerian subdigraph of D, we conclude that $\text{diff}(D) = \text{diff}(D_1) \times \text{diff}(D_2) \ne 0$. Hence D is an f-AT orientation of G. ∎

Discharging method is often used to find the existence of a subset X for which we can apply Lemma 3.8.1. This is explained in the proof of Theorem 3.8.2, which is given in [46]. For a positive integer l, we denote by $\mathcal{P}_{4,l}$ the family of planar graphs with no cycles of length 4 and l.

Theorem 3.8.2 *Every graph $G \in \mathcal{P}_{4,5}$ has a matching M such that $AT(G - M) \le 3$.*

For the purpose of using induction, instead of proving Theorem 3.8.2 directly, we shall prove a stronger and more technical result.

Definition 3.8.3 *Assume that G is a plane graph and v_0 is a vertex on the boundary of G. A valid matching of (G, v_0) is a matching M of G that does not cover v_0.*

Definition 3.8.4 *Let G be a plane graph and v_0 be a vertex on the boundary of G. An orientation D of G is good with respect to v_0, if D is AT with $\Delta_D^+(v) < 3$ and $d_D^+(v_0) = 0$.*

We shall prove the following result, which implies Theorem 3.8.2.

Theorem 3.8.5 *Assume that $G \in \mathcal{P}_{4,5}$, v_0 is a vertex on the boundary of G. Then (G, v_0) has a valid matching M such that there is a good orientation D of $G - M$ with respect to v_0.*

Assume that Theorem 3.8.5 is not true and G is a counterexample with $|V(G)| + |E(G)|$ minimum among all counterexamples. Let f_0 denote the outer face of G.

Lemma 3.8.6 *G is 2-connected. Moreover, $d(v) \ge 3$ for all $v \in V(G)\backslash\{v_0\}$.*

Proof. Assume that G has a cut vertex. We may choose a block B of G that contains a unique cut vertex z^* and does not contain

v_0. Let $G_1 = G - (B - \{z^*\})$. By the minimality of G, (G_1, v_0) has a valid matching M_1 and there is a good orientation D_1 of $G_1 - M_1$, and (B, z^*) has a valid matching M_2 and there is a good orientation D_2 of $B - M_2$. Let $M = M_1 \cup M_2$ and $D = D_1 \cup D_2$. Applying Lemma 3.8.1 (with $X = V(B) - \{z^*\}$), D is an AT-orientation. So M is a valid matching of (G, v_0), and $G - M$ has a good orientation, a contradiction.

If $d_G(v) \le 2$, then $G' = G - v$ has a valid matching M such that $G' - M$ has a good orientation D'. Extending D' to an orientation D of G in which v is a source vertex, it is obvious that D is a good orientation of $G - M$. ∎

Lemma 3.8.7 *G does not contain two adjacent 3-vertices u and v, with $u \ne v_0$ and $v \ne v_0$.*

Proof. Assume to the contrary that $uv \in E(G)$ with $d(u) = d(v) = 3$ and $u \ne v_0$ and $v \ne v_0$. Let $G^* = G - \{u, v\}$. Then (G^*, v_0) has a valid matching M^* such that there exists a good orientation D^* of $G^* - M^*$. Let $M = M^* \cup \{uv\}$. Then M is a valid matching of (G, v_0). Extend D^* to an orientation D of $G - M$ in which u, v are sources. Then D is a good orientation of $G - M$. ∎

For positive integers a, b, c, an (a, b, c)-triangle is a triangle in G whose three vertices have degrees a, b, c, respectively. An a-vertex (respectively, an a^+-vertex or an a^--vertex) is a vertex in G of degree a (respectively, at least a or at most a).

Definition 3.8.8 *A 3-face $f = [uvw]$ with $v_0 \notin \{u, v, w\}$ is called a minor triangle if f is a $(3, 4, 4)$-face. A 3-vertex $v \ne v_0$ is called a minor 3-vertex if v is incident to a triangle.*

Definition 3.8.9 *A triangle chain in G of length k is a subgraph of $G - v_0$ consisting of vertices $w_1, w_2, \ldots, w_{k+1}, u_1, u_2, \ldots, u_k$ in which $[w_i w_{i+1} u_i]$ is a $(4, 4, 4)$-face for $i = 1, 2, \ldots, k$. We denote by T_i the triangle $[w_i w_{i+1} u_i]$ and denote such a triangle chain by $T_1 T_2 \ldots T_k$.*

Observe that a single $[444]$-triangle is also a triangle chain. We say that a triangle T intersects a triangle chain $T_1 T_2 \ldots T_k$, if T has one common vertex with T_1.

Lemma 3.8.10 *If a minor triangle T_0 intersects a triangle chain $T_1 T_2 \ldots T_k$, then no vertex of T_k is adjacent to a 3-vertex, except possibly v_0; and no vertex of a minor triangle is adjacent to a 3-vertex, except possibly v_0.*

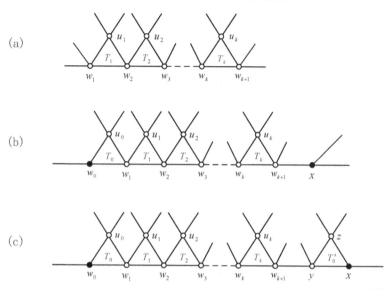

(a)

(b)

(c)

FIGURE 3.14: (a) A triangle chain. (b) A triangle chain that intersects a minor triangle and adjacent to a 3-vertex. (c) A triangle chain that intersects a minor triangle and has distance 1 to another minor triangle.

Proof. Assume to the contrary that $T_0 = [w_0 w_1 u_0]$ is a minor triangle that intersects a triangle chain $T_1 T_2 \ldots T_k$, with $T_i = [w_i w_{i+1} u_i]$ ($1 \leq i \leq k$), and w_{k+1} has a 3-neighbour x, as depicted in Figure 3.14(b). Assume that w_0 is a 3-vertex. Let $X = \cup_{i=0}^k V(T_i) \cup \{x\}$ and $G' = G - X$. By the minimality of G, (G', v_0) contains a valid matching M' and there is a good orientation D' of $G' - M'$.

Let $M = M' \cup \{w_0 u_0, w_1 u_1, \ldots, w_k u_k, w_{k+1} x\}$. Then M is a valid matching of (G, v_0). Let D be an orientation of $G - M$ obtained from D' by adding arcs (w_i, w_{i+1}) and (w_{i+1}, u_i) for $i = 0, 1, \ldots, k$, and all the edges between X and $V \backslash X$ are oriented from X to $V \backslash X$. Since $D[X]$ is acyclic, $D[X]$ is AT. By Lemma 3.8.1, D is AT. It is easy to see that $\Delta_D^+(v) < 3$ and $d_D^+(v_0) = 0$. Thus D is a good orientation of $G - M$. ∎

Lemma 3.8.11 *If a triangle chain $T_1 T_2 \ldots T_k$ intersects a minor triangle T_0, then the distance between T_k and another minor triangle is at least 2, and the distance between two minor triangles is at least 2.*

Proof. Assume to the contrary that $T_1 T_2 \ldots T_k$ with $T_i = [w_i w_{i+1} u_i]$ $(1 \le i \le k)$ is a triangle chain that intersects a minor triangle $T_0 = [w_0 w_1 u_0]$, and the distance between T_k and another minor triangle $T_0' = [xyz]$ with $d(x) = 3$ is less than 2. Combining with Lemma 3.8.10, assume $w_{k+1} y$ is a $(4,4)$-edge connecting T_k and T_0', as in Figure 3.14(c). Let $X = \cup_{i=0}^{k} V(T_i) \cup V(T_0')$ and $G' = G - X$. Then (G', v_0) has a valid matching M' and there is a good orientation D' of $G' - M'$.

Let $M = M' \cup \{w_0 u_0, w_1 u_1, \ldots, w_k u_k, w_{k+1} y, xz\}$. Then M is a valid matching of (G, v_0). Let D be an orientation of $G - M$ obtained from D' by adding arcs (x, y), (y, z), (w_i, w_{i+1}) and (w_{i+1}, u_i) for $i = 0, 1, \ldots, k$, and all the edges between X and $V \backslash X$ are oriented from X to $V \backslash X$. Obviously, $D[X]$ is acyclic, so $D[X]$ is AT. By Lemma 3.8.1, D is a good orientation of $G - M$, a contradiction. ∎

Lemma 3.8.12 *Assume that f is a 6-face of G which is adjacent to five triangles, and none of the vertices in these triangles is v_0. If f has one 3-vertex, then there is at least one 5^+-vertex on the five triangles.*

Proof. Let $f = [v_1 v_2 v_3 v_4 v_5 v_6]$, v_1 be a 3-vertex and $T_i = [v_i v_{i+1} u_i]$ $(i = 1, 2, \ldots, 5)$ be the five triangles (see Figure 3.15(a)). Assume to the contrary that there is no 5^+-vertex on T_i for $i = 1, 2, 3, 4, 5$. By Lemma 3.8.10, we may assume all v_{i+1} and u_i are 4-vertices for $i = 1, 2, \ldots, 5$. Let $X = \cup_{i=1}^{5} V(T_i)$ and $G' = G - X$. Then (G', v_0) has a valid matching M' and there is a good orientation D' of $G' - M'$.

Let $M = M' \cup \{v_1 u_1, v_2 u_2, \ldots, v_5 u_5\}$. Then M is a valid matching of (G, v_0). Let D be the orientation of $G - M$ obtained from D' by adding arcs (v_{i+1}, u_i), (v_i, v_{i+1}) for $i = 1, \ldots, 5$ and (v_1, v_6) and all the edges between X and $V \backslash X$ are oriented from X to $V \backslash X$ (see Figure 3.15(b)). Clearly, $\Delta_D^+(v) < 3$ and D is AT by Lemma 3.1.16, a contradiction. ∎

Now we use the discharging method to derive a contradiction. Let the *initial charge* ch be defined as $ch(x) = d(x) - 4$ for $x \in V(G) \cup F(G)$, where $F(G)$ is the set of faces of G. Applying the equalities $\sum_{v \in V(G)} d(v) = 2|E(G)| = \sum_{f \in F(G)} d(f)$ and Euler's formula $|V(G)| - |E(G)| + |F(G)| = 2$, we conclude that

$$\sum_{x \in V(G) \cup F(G)} ch(x) = -8.$$

We shall describe a set of rules that transfer charge from some vertices and faces to other vertices and faces. In the discharging procedure, $ch(x \to y)$ denotes the charge discharged from an element x to another element y, $ch(x \to)$ and $ch(\to x)$ denote the charge totally discharged from or to x, respectively. The *final charge* $ch^*(x)$ of $x \in V(G) \cup F(G)$ is defined as $ch^*(x) = ch(x) - ch(x \to) + ch(\to x)$. By applying appropriate discharging rules, we shall arrive at a final charge that $ch^*(x) \geq 0$ for all $x \in V(G) \cup F(G) \setminus \{v_0, f_0\}$, and $ch^*(v_0) + ch^*(f_0) > -8$. As the total charge does not change in the discharging process, this would be a contradiction.

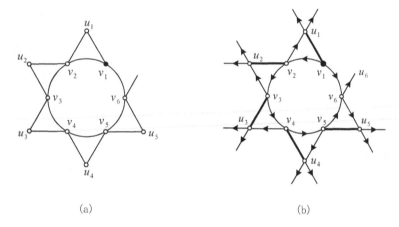

(a)   (b)

FIGURE 3.15: The configuration in Lemma 3.8.12.

The discharging rules are as follows:

R1 Assume that $f \neq f_0$ is a 3-face. Then each face adjacent to f transfers $\frac{1}{3}$ charge to f.

R2 Assume that $v \neq v_0$ is 3-vertex. If v is contained in a triangle, then each of the other two faces incident to v transfers $\frac{1}{2}$ charge to v; otherwise each face incident to v transfers $\frac{1}{3}$ charge to v.

R3 Assume that $u \neq v_0$ is a 5^+-vertex and $f \neq f_0$ is a 6-face.

    (i) If f is incident to u and adjacent to two triangles that are incident to u, then u transfers $\frac{1}{3}$ charge to f.

    (ii) If f is incident to u and adjacent to exactly one triangle which is incident to u, then u transfers $\frac{1}{6}$ charge to f.

(iii) If f is not incident to u but adjacent to a triangle which is incident to u, then u transfers $\frac{1}{6}$ charge to f.

R4 f_0 transfers $\frac{1}{3}$ charge to each adjacent triangle, and $\frac{1}{2}$ charge to each incident 3-vertex $v \neq v_0$. v_0 transfers $\frac{1}{3}$ charge to each 6-face $f \neq f_0$ which is either incident to v_0, or is not incident to v_0 but adjacent to a triangle T which is incident to v_0.

Claim 3.8.13 *If a 6-face f has three 3-vertices other than v_0 and is adjacent to three triangles, then $ch(\to f) \geq \frac{1}{2}$.*

Proof. Assume that $f = [v_1 v_2 v_3 v_4 v_5 v_6]$. By Lemma 3.8.7, we may assume that v_1, v_3 and v_5 are the three 3-vertices other than v_0. Then each of v_1, v_3, v_5 is incident to at most one triangle. Hence at most two of the three triangles adjacent to f intersect each other. Thus we may assume that the three triangles adjacent to f are either T_1, T_2, T_4, or T_1, T_3, T_5, where $T_i = [v_i v_{i+1} u_i]$.

Case 1 The three triangles incident to f are T_1, T_2, T_4.

If v_0 is a vertex of f or T_1, T_2 or T_4, then v_0 transfers $\frac{1}{3}$ charge to f by R4. By Lemma 3.8.10, at least one of the three triangles has a 5^+-vertex $v \neq v_0$ which sends at least $\frac{1}{6}$ charge to f. So $ch(\to f) \geq \frac{1}{3} + \frac{1}{6} = \frac{1}{2}$.

Assume that v_0 is not a vertex of f, T_1, T_2 or T_4. By Lemma 3.8.10, either v_2 is a 5^+-vertex or both of u_1 and u_2 are 5^+-vertices. In both cases, f receives $\frac{1}{3}$ charge in total from v_2, u_1 and u_2. Moreover, by Lemma 3.8.10, either v_4 or u_4 is a 5^+-vertex, which transfers $\frac{1}{6}$ charge to f. Hence, $ch(\to f) \geq \frac{1}{3} + \frac{1}{6} = \frac{1}{2}$.

Case 2 The three triangles incident to f are T_1, T_3, T_5.

By Lemma 3.8.10, each of these three triangles has a 5^+-vertex transferring $\frac{1}{6}$ charge to f. Thus, $ch(\to f) \geq \frac{1}{2}$. ∎

Claim 3.8.14 *If a 6-face f has two 3-vertices other than v_0 and is adjacent to four triangles, then $ch(\to f) \geq \frac{1}{3}$.*

Proof. Assume that $f = [v_1 v_2 v_3 v_4 v_5 v_6]$. If edge $v_i v_{i+1}$ is contained in a triangle, then we denote the triangle by $T_i = [v_i v_{i+1} u_i]$. If v_0 is a vertex of f or T_i, then v_0 transfers $\frac{1}{3}$ charge to f by R4. Suppose v_0 is not a vertex of f or T_i. By Lemma 3.8.7, we may assume that either v_1 and v_4 or v_1 and v_3 are the two 3-vertices. Analysing all the distribution of the four triangles, we need to consider the following five cases:

Case 1 The four triangles incident to f are T_1, T_2, T_3, T_5 while the two 3-vertices are v_1 and v_4.

If at least one of v_2 and v_3 is a 5^+-vertex, then by R3(i), $ch(\to f) \geq \frac{1}{3}$. Assume that both $d(v_2)$ and $d(v_3)$ are 4-vertices. By Lemma 3.8.10, at least two of u_1, u_2 and u_3 are 5^+-vertices each of which transfers $\frac{1}{6}$ charge to f. That is, $ch(\to f) \geq \frac{1}{3}$.

Case 2 The four triangles incident to f are T_1, T_2, T_4, T_5 while the two 3-vertices are v_1 and v_4.

By Lemma 3.8.10 and R3, at least one of u_1, u_2, v_2 and v_3 is a 5^+-vertex transferring at least $\frac{1}{6}$ charge to f. By symmetry, there is another such vertex in u_4, u_5, v_5 and v_6. Thus, we are done.

Case 3 The four triangles incident to f are T_1, T_2, T_4, T_5 while the two 3-vertices are v_1 and v_3.

If v_2 is a 5^+-vertex, then v_2 transfers $\frac{1}{3}$ charge to f by R3(i). Assume that v_2 is a 4-vertex. From Lemma 3.8.10, both u_1 and u_2 are 5^+-vertices each of which transfers $\frac{1}{6}$ charge to f.

Case 4 The four triangles incident to f are T_1, T_3, T_4, T_5 while the two 3-vertices are v_1 and v_3.

By Lemma 3.8.10, at least one of v_2 and u_1 is a 5^+-vertex transferring $\frac{1}{6}$ charge to f. Moreover, using Lemma 3.8.10 again, at least one of v_4, v_5, v_6, u_3, u_4 and u_5 is a 5^+-vertex transferring at least $\frac{1}{6}$ to f. Consequently, $ch(\to f) \geq \frac{1}{3}$.

Case 5 The four triangles incident to f are T_3, T_4, T_5, T_6 while the two 3-vertices are v_1 and v_3.

If one of v_4, v_5 and v_6 is a 5^+-vertex, then such a 5^+-vertex sends $\frac{1}{3}$ charge to f by R3(i). Assume that all of v_4, v_5 and v_6 are 4-vertices. Then at least two of u_3, u_4, u_5 and u_6 are 5^+-vertices each sending $\frac{1}{6}$ to f. Otherwise, it will contradict Lemma 3.8.10 or Lemma 3.8.11. Again $ch(\to f) \geq \frac{1}{3}$. ∎

Now we check the final charge of vertices $v \neq v_0$.

If v is a 3-vertex, then by R2, v receives 1 from incident 6^+-faces. So $ch^*(v) = ch(v) - ch(v \to) + ch(\to v) = -1 - 0 + 1 = 0$.

If v is a 4-vertex, no charge is sent out or received. So $ch^*(v) = ch(v) = 0$.

If v is a 5^+-vertex, then by the rules, v only transfers charge to faces when it is incident with a triangle. Assume that v is incident with t triangles, then $0 < t \le \lfloor \frac{d(v)}{2} \rfloor$. Let T be a triangle incident with v, f_1 be a 6-face not incident to v but adjacent to T, and f_2, f_3 be 6-faces incident to v and adjacent to T. Note that, by R3, v transfers $\frac{1}{2}$ charge to those 6-faces around T. This means that v sends out at most $\frac{1}{2}t$ charge. So we have $ch^*(v) = ch(v) - ch(v \to)$ $\ge d(v) - 4 - \frac{1}{2}t \ge d(v) - 4 - \frac{1}{2} \times \lfloor \frac{d(v)}{2} \rfloor \ge 0$.

Next we check the final charge of faces $f \ne f_0$.

If f is a 3-face, then R1 guarantees $ch^*(f) \ge 0$. Assume that f is a 6-face. Note that if f has no minor 3-vertex, then f sends out at most $\frac{1}{3} \times 6 = 2$ to others and hence $ch^*(f) \ge 0$. Assume that f has s minor 3-vertices for $1 \le s \le 3$. When $s = 3$, f is adjacent to at most three triangles each of which transfers $\frac{1}{3}$ charge from f. By Claim 3.8.13, we have $ch^*(f) = d(f) - 4 - ch(f \to) + ch(\to f)$ $\ge 2 - (\frac{1}{2} \times 3 + \frac{1}{3} \times 3) + \frac{1}{2} = 0$. When $s = 2$, we may assume that f has exactly two 3-vertices and is adjacent to four triangles. Otherwise $ch(f \to) \le \frac{1}{2} \times 2 + \frac{1}{3} \times 3 = 2$. Thus, by Claim 3.8.14, we have $ch^*(f) = d(f) - 4 - ch(f \to) + ch(\to f) \ge 2 - (\frac{1}{2} \times 2 + \frac{1}{3} \times 4) + \frac{1}{3} = 0$. Next we consider that $s = 1$. We may assume that f has exactly one 3-vertex and is adjacent to five triangles. Otherwise $ch(f \to) \le \frac{1}{2} + \frac{1}{3} \times 4 = \frac{11}{6} < 2$, and we are done. Thus, $ch(f \to) = \frac{1}{2} + \frac{1}{3} \times 5 = \frac{13}{6}$. On the other hand, by Lemma 3.8.12, there is at least one 5^+-vertex transferring $\frac{1}{6}$ charge to f on the five triangles. This gives that $ch^*(f) \ge 2 - \frac{13}{6} + \frac{1}{6} = 0$.

If f is a 7^+-face with s 3-vertices, then f is adjacent to at most $(d(f) - s)$ triangles. Obviously, $0 \le s \le \lfloor \frac{d(f)}{2} \rfloor$. Hence $ch^*(f) = d(f) - 4 - [\frac{1}{2} \times s + \frac{1}{3} \times (d(f) - s)] = \frac{2}{3}d(f) - \frac{1}{6}s - 4 > 0$.

Finally we check the charges on f_0 and v_0.

By R4, v_0 transfers at most $(d(v_0) - 1) \times \frac{1}{3}$ charge to others. That is, $ch^*(v_0) \ge d(v_0) - 4 - (d(v_0) - 1) \times \frac{1}{3} = \frac{2}{3}d(v_0) - \frac{11}{3} \ge -\frac{7}{3}$, as $d(v_0) \ge 2$.

The face f_0 is incident with at most $\lfloor \frac{d(f_0)}{2} \rfloor$ 3-vertices and each receiving $\frac{1}{2}$ charge from f_0, and f_0 is adjacent to at most $d(f_0)$ triangles and each receiving $\frac{1}{3}$ charge from f_0. Therefore $ch^*(f_0) \ge d(f_0) - 4 - \frac{1}{2}\lfloor \frac{d(f_0)}{2} \rfloor - \frac{1}{3}d(f_0) \ge \frac{5}{12}d(f_0) - 4 \ge -\frac{11}{4}$. Consequently, the sum of the final charge of all vertices and faces is

$$\sum_{x \in V \cup F} ch^*(x) \ge ch^*(v_0) + ch^*(f_0) \ge -\frac{61}{12},$$

contrary to the fact that $\sum_{x \in V \cup F} ch^*(x) = \sum_{x \in V \cup F} ch(x) = -8$. This completes the proof of Theorem 3.8.5. ∎

Similar argument is used in [46] to show that for $l \in \{6, 7\}$, every graph $G \in \mathcal{P}_{4,l}$ has a matching M such that $G - M$ has Alon–Tarsi number at most 3.

3.9 Hypergraph colouring

Assume that $k \geq 2$ and $H = (V, E)$ is a k-uniform hypergraph with vertex set $V = \{v_1, v_2, \ldots, v_n\}$. Each hyperedge $e \in E$ of H is a set of k vertices of H. Assume that $e = \{v_{i_1}, v_{i_2}, \ldots, v_{i_k}\}$, where $i_1 < i_2 < \ldots < i_k$. Then v_{i_j} is the jth vertex of e. For convenience, we write a hyperedge $e = \{v_{i_1}, v_{i_2}, \ldots, v_{i_k}\}$ as an injective mapping $e : [k] \to [n]$ defined as $e(j) = i$ if the jth vertex of e is v_i.

A proper colouring of H is a mapping ϕ which assigns to each vertex v_i a colour $\phi(v_i)$ such that no hyperedge is monochromatic, i.e., each hyperedge e has at least two vertices that are coloured by distinct colours. Ramamurti and West [52] extended the application of CNS to colouring of uniform hypergraphs. As for graphs, each vertex v_i is associated with a variable x_i. Let θ be a primitive k-th root of unity. For a hyperedge e, let

$$h_e(\mathbf{x}) = \theta x_{e(1)} + \theta^2 x_{e(2)} + \ldots + \theta^k x_{e(k)}.$$

Note that $\theta^k = \theta^0 = 1$ and $\theta + \theta^2 + \ldots + \theta^k = 0$.

The *hypergraph polynomial* f_H for H is defined as

$$f_H(\mathbf{x}) = \prod_{e \in E(H)} h_e(\mathbf{x}).$$

Assume that ϕ is a mapping from V to $\mathbb{R}$. For a hyperedge e, if $\phi(v_{e(j)}) = c$, a constant for $j = 1, 2, \ldots, k$, then

$$h_e(\phi) = h_e(\phi(x_1), \phi(x_2), \ldots, \phi(x_k)) = c(\theta + \theta^2 + \ldots + \theta^k) = 0.$$

Note that this is a natural generalization of what had been defined for graphs for which $k = 2$ and $\theta = -1 =$ the primitive square root of unity. Hence if $f_H(\phi) \neq 0$, then each hyperedge uses at least two distinct colours and ϕ is a proper colouring of H. However, the converse is not true in general. For example, if $k = 2q \geq 4$ is even, then for e, let $\phi(v_{e(q)}) = \phi(v_{e(k)}) = c \neq 0$ and $\phi(v_{e(j)}) = 0$ for $j \neq k, q$, then $h_e(\phi) = 0$. However, if k is a prime, then the following holds ([6], p. 405):

Lemma 3.9.1 *If k is a prime number, and θ is a primitive k-th root of unity and $a_1, a_2, \ldots, a_k$ are rational, then $\sum_{i=0}^{k-1} a_i \theta^i = 0$ if and only if $a_0 = a_1 = \ldots = a_{k-1}$.*

Thus if k is a prime, then $f_H(\phi) \neq 0$ if and only if ϕ is a proper colouring of H.

In the remainder of this section, we assume that k is a prime and $H = (V, E)$ is a k-uniform hypergraph.

Definition 3.9.2 *An orientation of H is a mapping σ which assigns to each edge e a vertex of e as its* source vertex. *If v is the source vertex of e, then e is called an* out-edge *at v.*

Since each hyperedge e is an injective mapping from $[k]$ to $[n]$, an orientation of H can be encoded as a mapping $\sigma : E \to \{1, 2, \ldots, k\}$, where $\sigma(e) = j$ means that the jth vertex of e is the source vertex of e. In other words, the source vertex of e under the orientation σ is $v_{e(\sigma(e))}$.

We define the weight of e under the orientation σ as

$$w_\sigma(e) = \theta^{\sigma(e)} x_{e(\sigma(e))},$$

and let

$$w(\sigma) = \prod_{e \in E} w_\sigma(e).$$

For $i = 1, 2, \ldots, n$, let

$$E_\sigma(v_i) = \{e \in E(v) : e(\sigma(e)) = i\},$$

i.e., $E_\sigma(v_i)$ is the set of out-edges at v_i. The number $d_i = |E_\sigma(v_i)|$ is called the *out-degree* of v_i, and the sequence $(d_1, d_2, \ldots, d_n)$ is called the *out-degree sequence of σ* and is denoted by $\vec{d}(\sigma)$. Let

$$m(\sigma) = \sum_{e \in E} \sigma(e) \pmod{k}.$$

It follows from the definition that if $\vec{d}(\sigma) = (d_1, d_2, \ldots, d_n)$, then

$$w(\sigma) = \theta^{m(\sigma)} \prod_{i=1}^{n} x_i^{d_i}.$$

Note that if $k = 2$, then H is a graph, and $m(\sigma) = 1$ or -1, according to whether σ has an even or odd number of reverse arcs.

For an out-degree sequence $\vec{d} = (d_1, d_2, \ldots, d_n)$, for $i \in \{0, 1, \ldots, k - 1\}$, let $a_{\vec{d},i}$ be the number of orientations σ with $\vec{d}(\sigma) = \vec{d}$ and $m(\sigma) = i$. The same proof as Theorem 3.1.9 shows that

$$f_H(\mathbf{x}) = \sum_{\vec{d}=(d_1,d_2,\ldots,d_n),\sum_{i=1}^{n} d_i=|E|} \left(\sum_{j=0}^{k-1} a_{\vec{d},j} \theta^j \right) \prod_{i=1}^{n} x_i^{d_i}.$$

Thus a monomial $\prod_{i=1}^{n} x_i^{d_i}$ is vanishing if and only if $\sum_{j=0}^{k-1} a_{\vec{d},j} \theta^j = 0$. By Lemma 3.9.1, $\sum_{j=0}^{k-1} a_{\vec{d},j} \theta^j = 0$ if and only if there is a constant a such that $a_{\vec{d},j} = a$ for $j = 0, 1, \ldots, k - 1$. In particular, we have the following corollary.

Corollary 3.9.3 *If the total number of orientations σ of H with $\vec{d}(\sigma) = (d_1, d_2, \ldots, d_n)$ is not a multiple of k, then $\prod_{i=1}^{n} x_i^{d_i}$ is a non-vanishing monomial of f_H.*

Assume that $k = 2$ and σ is an orientation of H with out-degree sequence $\vec{d}$. Then all other orientations σ' of H with the same out-degree sequence can be obtained from σ by reversing the orientation of the edges of an Eulerian subdigraph of σ. In some sense, the same holds for k-uniform hypergraphs with $k \geq 3$. To describe the corresponding result for the $k \geq 3$ case, we need a notion that corresponds to the operation of 'reversing the orientation of edges in an Eulerian subdigraph of σ'.

For the $k = 2$ case, using the notation introduced in this section, reversing the direction of an edge e of H is equivalent to increasing the value of $\sigma(e)$ by 1 (modulo 2). To be precise, if E' is a subset E, and σ' is obtained from σ by reversing the orientations of edges in E', then σ' is defined as

$$\sigma'(e) = \begin{cases} \sigma(e) + 1 \pmod{2}, & \text{if } e \in E', \\ \sigma(e), & \text{otherwise.} \end{cases}$$

Alternatively, let $\tau : E \to \{0, 1\}$ be defined as $\tau(e) = 1$ if $e \in E'$ and $\tau(e) = 0$ otherwise. Then $\sigma' = \sigma + \tau$, i.e., $\sigma'(e) = \sigma(e) + \tau(e)$ (mod 2).

For $k \geq 3$, the operation of 'reversing the orientation of some edges' can be defined in the same way. A *shift function* is a mapping

$\tau : E \to \{0, 1, \ldots, k - 1\}$. The orientation $\sigma' = \sigma + \tau$ defined as $\sigma'(e) = \sigma(e) + \tau(e) \pmod{k}$ is said to be the τ-shift of σ. A τ-shift of σ can be viewed as a modification of the orientation σ. We need to put restrictions on τ so that the τ-shift of σ has the same out-degree sequence as σ.

In case $k = 2$, then $\sigma' = \sigma + \tau$ has the same out-degree sequence as σ if and only if the set $\{e : \tau(e) = 1\}$ is an Eulerian subdigraph of σ.

Assume that e is an edge of G and v is a vertex of e. Let $i_e(v)$ be the position of v in e, i.e., v is the $i_e(v)$-th vertex of e. Then

$$\sigma(e) - i_e(v)$$

is the difference between the position of the source vertex of e and the position of v in e. In particular, $\sigma(e) - i_e(v) = 0$ if and only if v is the source vertex of e.

Assume that $\sigma' = \sigma + \tau$.

Let $E_{\sigma,\tau}(v) = \{e \in E(v) : \tau(e) = i_e(v) - \sigma(e)\}$. Then σ' and σ have the same out-degree sequence if and only if for each vertex v,

$$|E_\sigma(v)| = |E_{\sigma,\tau}(v)|.$$

Let $E_\tau^* = \{e \in E(v) : \tau(e) = 0\}$. Then E_τ^* consists of those edges e whose orientations are not changed. The set $E_\sigma(v) - E_\tau^*(v)$ consists of those edges that are out-edges at v in σ, but not out-edges at v in σ'; and the set $E_{\sigma,\tau}(v) - E_\tau^*(v)$ consists of those edges that are out-edges at v in σ', but not out-edges at v in σ.

We call a shift function τ *balanced for* σ if for each vertex v, $|E_\sigma(v) - E_\tau^*(v)| = |E_{\sigma,\tau}(v) - E_\tau^*(v)|$. Then balanced shift functions play the role of Eulerian subdigraphs in the $k = 2$ case, and every orientation σ' with $\vec{d}(\sigma') = \vec{d}(\sigma)$ is obtained from σ by adding a balanced shift function, i.e., $\sigma' = \sigma + \tau$ for some balanced shift function τ for σ. Moreover, let $m(\tau) = \sum_{e \in E} \tau(e)$. Then $w(\sigma + \tau) = \theta^{m(\tau)} w(\sigma)$. Assume that $\vec{d}(\sigma) = (d_1, d_2, \ldots, d_n)$. Then the coefficient of the monomial $\prod_{i=1}^{n} x_i^{d_i}$ in the expansion of f_H is

$$\sum_\tau \theta^{m(\tau)} \theta^{m(\sigma)},$$

where the summation is over all balanced shift functions τ of σ.

We say that a hypergraph H is f-*Alon–Tarsi choosable*, f-AT for short, if f_H has a non-vanishing monomial $\prod_{i=1}^{n} x_i^{t_i}$ with $t_i < f(v_i)$. Hence we have the following corollary.

Corollary 3.9.4 *Assume that σ is an orientation of a k-uniform hypergraph H with out-degree sequence $\vec{d}(\sigma) = (d_1, d_2, \ldots, d_n)$, and L is a list assignment which assigns to each vertex v_i a set $L(v_i)$ of $d_i + 1$ integers as permissible colours. If $\sum_\tau \theta^{m(\tau)}$ over all balanced shift functions τ of σ is non-zero, then H is L-colourable.*

Corollary 3.9.5 *Assume that σ is an orientation of H with out-degree sequence $\vec{d}(\sigma) = (d_1, d_2, \ldots, d_n)$. If the total number of balanced shift functions for σ is not a multiple of k, then $\prod_{i=1}^{n} x_i^{d_i}$ is a non-vanishing monomial of f_H. Hence with $f(v_i) = d_i + 1$, H is f-AT.*

Observe that for any orientation σ, τ_0 defined as $\tau_0(e) = 0$ for each hyperedge e is a trivial balanced shift function for σ.

Example 3.9.6 *The Fano plane F as given in Figure 3.16 is a 3-uniform hypergraph with 7 vertices $\{1, 2, \ldots, 7\}$ and 7 edges $\{124, 235, 136, 157, 267, 347, 456\}$. Let σ be the orientation of F given as $\{1\underline{2}4, \underline{2}35, 1\underline{3}6, 1\underline{5}7, \underline{2}67, \underline{3}47, \underline{4}56\}$ (where the underlined vertex in each edge is the source vertex). Then there is only one non-trivial balanced shift function τ for σ defined as $\tau(124) = \tau(235) = 1, \tau(136) = 2$ and $\tau(e) = 0$ for all the other edges e.*

It is easy to verify that τ defined above is indeed a balanced shift function for σ. Indeed, $\sigma + \tau$ is the orientation $\{1\underline{2}4, 2\underline{3}5, \underline{1}36, 1\underline{5}7, \underline{2}67, \underline{3}47, \underline{4}56\}$, which has the same out-degree sequence as σ. On the other hand, if τ is a balanced shift function for σ, then since the vertices $5, 6, 7$ have out-degree 0, we have $\tau(e) \neq 1$ for $e \in \{136, 157, 267, 456\}$ and $\tau(e) \neq 2$ for $e \in \{235, 157, 267, 347, 456\}$. Hence $\tau(456) = 0$. This in turn implies that $\tau(347) \neq 1$ and $\tau(124) \neq 2$. Thus $\tau(e) = 1$ only if $e \in \{124, 235\}$ and $\tau(e) = 2$ only if $e = 136$. It is straightforward to verify that $\tau(124) = 1 \Rightarrow \tau(235) = 1 \Rightarrow \tau(136) = 2 \Rightarrow \tau(124) = 1$.

As the total number of balanced shift functions for σ is not a multiple of 3, and as each vertex of F has out-degree at most 2 in σ, we conclude that F is 3-choosable.

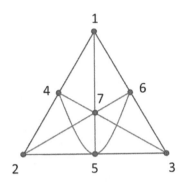

FIGURE 3.16: The Fano plane F.

3.10 Paintability of graphs

The Alon–Tarsi number $AT(G)$ of G is not only an upper bound for the choice number of G, but it is also an upper bound for the painter number of G, which is an on-line version of the choice number. The painter number of a graph is defined through a two-person game. Assume that G is a graph and $f : V(G) \to Z^+$ is a function which assigns to each vertex a positive integer.

Definition 3.10.1 *[53] The f-painting game on G is a two-person game defined as follows: The two players are Lister and Painter. Initially no vertex of G is coloured and each vertex v is given $f(v)$ tokens. In each round, Lister chooses a subset M of uncoloured vertices, and removes one token from each vertex in M. Painter chooses a subset I of M that is an independent set of G and colours vertices in I. If at the end of a certain round, there is an uncoloured vertex with no tokens left, then Lister wins the game. Otherwise, at the end of a certain round, all vertices of G are coloured, then Painter wins the game.*

Definition 3.10.2 *We say that G is f-paintable if Painter has a winning strategy for this game. We say that G is k-paintable if G is f-paintable for the constant function $f(v) = k$ for all $v \in V(G)$. The* painter number *$\chi_P(G)$ is defined by*

$$\chi_P(G) = \min\{k : G \text{ is } k\text{-paintable}\}.$$

Consider the case when Lister plays with a trivial strategy: On each round, Lister chooses the set of all uncoloured vertices. If $f(v) = k$ for all vertices v and Lister plays with this trivial strategy, then Painter can win the game if and only if G is k-colourable. With this trivial strategy of Lister, the game ends in at most k rounds: after k rounds, if there are still uncoloured vertices, then all the tokens of each uncoloured vertex are gone, and Lister wins the game. Otherwise all vertices are coloured and Painter wins. If G is k-colourable, then $V(G)$ is covered by k independent sets $I_1, I_2, \ldots, I_k$. Painter's strategy is to colour I_j in the jth round. Conversely, if Painter wins the game, then let I_j be the set of vertices coloured in the jth round. Then $I_1, I_2, \ldots, I_k$ are k independent sets that form a partition of $V(G)$. So G is k-colourable. This shows that $\chi_P(G) \geq \chi(G)$.

Next we observe that the painter number of G is actually an upper bound for the choice number of G.

Lemma 3.10.3 *For any graph G, $ch(G) \leq \chi_P(G)$.*

Proof. Assume that $\chi_P(G) = k$ and L is a k-list assignment of G. We shall prove that G is L-colourable.

Let $\bigcup_{v \in V(G)} L(v) = \{1, 2, \ldots, m\}$. We play the k-painting game on G. In the ith round, Lister chooses the set containing all uncoloured vertices from $\{v : i \in L(v)\}$.

As G is k-paintable, Painter has a winning strategy, and hence after m rounds, as all the tokens are gone, all vertices are coloured. The result is an L-colouring of G. $\blacksquare$

The graph shown in Figure 3.17 is the $\Theta_{2,2,4}$ graph. It is known that this graph is 2-choosable, however, it is not 2-paintable. Lister's winning strategy is as follows: In the first move, Lister marks v_1, v_2. Painter can colour either v_1 or v_2. Assume that Painter colours v_1. Lister will mark sets $\{v_2, v_3\}, \{v_3, v_4\}$, $\{v_4, u, w\}, \{v_0, u\}, \{v_0, w\}$ in the next five moves. In each of these moves, Painter is forced to colour vertices v_2, v_3, v_4, u in the next four moves (otherwise there will be an uncoloured vertex with no tokens left and Painter loses), and in the fifth move, Painter loses.

If Painter colours v_2 in the first move, then Lister marks sets $\{v_0, v_1\}, \{v_0, u, w\}, \{u_4, u\}, \{u_4, w\}$ in the next four moves and Painter will lose.

It is known [20] that the difference $\chi_P(G) - ch(G)$ can be arbitrarily large.

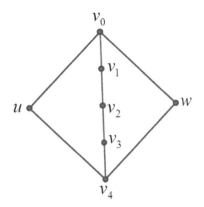

FIGURE 3.17: The graph $\Theta_{2,2,4}$.

Nevertheless, many upper bounds for the choice number of graphs are also upper bounds for their painter number. For example, we say that a graph G is *d-degenerate* if its vertices can be ordered as $v_1, v_2, \ldots, v_n$ so that each v_i has at most d neighbours v_j with $j < i$. Equivalently, G is d-degenerate if every subgraph of G contains a vertex of degree at most d. For example, every tree is 1-degenerate.

It is well-known that $ch(G) \leq d+1$ for any d-degenerate graph G. This upper bound holds for painter number as well, i.e., for any d-degenerate graph G, $\chi_P(G) \leq d+1$. Indeed, let $v_1, v_2, \ldots, v_n$ be an ordering of the vertices of G for which each v_i has at most d neighbours v_j with $j < i$. In each round, Painter colours vertices greedily: traversing from v_1 to v_n, Painter colours vertex v_i provided v_i is marked in this round and no earlier neighbour of v_i is coloured in this round. It is obvious that this is a winning strategy for Painter in the $(d+1)$-painting game on G.

It was proved by Schauz [53] that $AT(G)$ is also an upper bound for $\chi_P(G)$. The remainder of this section is devoted to the proof of this result.

For $f \in \mathbb{N}_0^G$ and $S \subseteq V(G)$, denote by $\mathcal{O}_G(f^S)$ the set of orientations D of G for which $d_D^+(v) = f(v)$ if $v \in V(G) \backslash S$ and $d_D^+(v) \geq f(v)$ if $v \in S$. Let $<$ be an arbitrary fixed ordering of vertices of G, and let the parity of an orientation D of G is the parity of the number of arcs (u, v) for which $u > v$. Let $\mathcal{O}_{e,G}(f^S)$ and $\mathcal{O}_{o,G}(f^S)$ be the sets of even and odd orientations in $\mathcal{O}_G(f^S)$, respectively. Let

$$\mathrm{diff}_G(f^S) = |\mathcal{O}_{e,G}(f^S)| - |\mathcal{O}_{o,G}(f^S)|.$$

In case $S = \emptyset$, then $\mathcal{O}_G(f^S), \mathcal{O}_{e,G}(f^S), \mathcal{O}_{o,G}(f^S), \mathrm{diff}_G(f^S)$ are denoted by $\mathcal{O}_G(f), \mathcal{O}_{e,G}(f), \mathcal{O}_{o,G}(f), \mathrm{diff}_G(f)$, respectively. Note that $\mathcal{O}_G(f)$ is the set of orientations D with $d_D^+(v) = f(v)$ for all v.

Lemma 3.10.4 *Assume that* $f \in \mathbb{N}_0^G$ *and* $S \subseteq V(G)$ *and* $\mathrm{diff}_G(f^S) \neq 0$. *Then the following hold:*

- *If* $\sum_{v \in V(G)} f(v) = |E(G)|$, *then for any subset* S' *of* $V(G)$, $\mathrm{diff}_G(f^{S'}) \neq 0$.

- *There exists* $g \in \mathbb{N}_0^G$ *such that* $g \geq f$, $g(v) = f(v)$ *for* $v \in V(G)\backslash S$, $\sum_{v \in V(G)} g(v) = |E(G)|$ *and* $\mathrm{diff}_G(g) \neq 0$.

Proof. Assume that $\sum_{v \in V(G)} f(v) = |E(G)|$. Then for any subset S' of $V(G)$, for any orientation $D \in \mathcal{O}_G(f^{S'})$, $|E(G)| = \sum_{v \in V(G)} f(v) \leq \sum_{v \in V(G)} d_D^+(v) = |E(G)|$. Hence $d_D^+(v) = f(v)$ for every vertex v. Thus $\mathcal{O}_G(f^{S'}) = \mathcal{O}_G(f^S)$ and $\mathrm{diff}_G(f^{S'}) = \mathrm{diff}_G(f^S) \neq 0$.

Let

$$\mathcal{F} = \left\{ g \in \mathbb{N}_0^G : g \geq f, g(v) = f(v) \forall v \in V(G)\backslash S, \right.$$

$$\left. \sum_{v \in V(G)} g(v) = |E(G)| \right\}.$$

Then $\mathcal{O}_G(f^S)$ is the disjoint union $\{\mathcal{O}_G(g) : g \in \mathcal{F}\}$ and $0 \neq |\mathcal{O}_{e,G}(f^S)| - |\mathcal{O}_{o,G}(f^S)| = \sum_{g \in \mathcal{F}} |\mathcal{O}_{e,G}(g)| - |\mathcal{O}_{o,G}(g)|$. Hence there exists $g \in \mathbb{N}_0^G$ such that $g \geq f$, $g(v) = f(v)$ for $v \in V(G)\backslash S$, $\sum_{v \in V(G)} = |E(G)|$ and $\mathrm{diff}_G(g) \neq 0$. ∎

Lemma 3.10.5 *If G has an edge $e = uv$ such that $f(u) = f(v) = 0$ and $u, v \in S$, then $\mathrm{diff}_G(f^S) = 0$.*

Proof. For any $D \in \mathcal{O}_G(f^S)$, let $\phi(D)$ be the orientation of G obtained from D by reversing the orientation of edge e. Then ϕ is a one-to-one correspondence from $\mathcal{O}_{e,G}(f^S)$ to $\mathcal{O}_{o,G}(f^S)$. So $\mathrm{diff}_G(f^S) = 0$. ∎

Lemma 3.10.6 *If $u \in S$ and $f(u) > 0$ and $\mathrm{diff}_G(f^S) \neq 0$, then $\mathrm{diff}_G(\hat{f}_u^S) \neq 0$ or $\mathrm{diff}_G(\hat{f}_u^{S-u}) \neq 0$.*

Proof. Assume that $u \in S$ and $f(u) > 0$. For $D \in \mathcal{O}_G(\hat{f}_u^S)$, either $d_D^+(u) \geq f(u)$ or $d_D^+(u) = f(u) - 1$. In the former case, $D \in \mathcal{O}_G(f^S)$. In the latter case, $D \in \mathcal{O}_G(\hat{f}_u^{S-u})$. It is obvious that $\mathcal{O}_G(f^S) \cup \mathcal{O}_G(\hat{f}_u^{S-u}) \subseteq \mathcal{O}_G(\hat{f}_u^S)$. Thus

$$\mathcal{O}_G(\hat{f}_u^S) = \mathcal{O}_G(f^S) \cup \mathcal{O}_G(\hat{f}_u^{S-u}).$$

Thus if $\mathrm{diff}_G(f^S) \neq 0$, then $\mathrm{diff}_G(\hat{f}^S) \neq 0$ or $\mathrm{diff}_G(\hat{f}^{S-u}) \neq 0$. ∎

Lemma 3.10.7 *Assume that $f \in \mathbb{N}_0^G$, $S \subseteq V(G)$, and $\mathrm{diff}_G(f^S) \neq 0$. Then there exist $R \subseteq S$, $g \in \mathbb{N}_0^G$ for which the following hold:*

1. $g \leq f$ and $g(v) < f(v)$ for any $v \in S \backslash R$.

2. $\mathrm{diff}_G(g^R) \neq 0$.

3. $g(v) = 0$ for every $v \in R$.

Proof. We prove the lemma by induction on $\sum_{v \in S} f(v)$.

If $f(v) = 0$ for all $v \in S$, then let $R = S$ and $g = f$ and we are done. So assume that $f(u) > 0$ for some $u \in S$. By Lemma 3.10.6, $\mathrm{diff}_G(\hat{f}_u^S) \neq 0$ or $\mathrm{diff}_G(\hat{f}_u^{S-u}) \neq 0$.

Let $f' = \hat{f}_u$. In the first case, let $S' = S$ and in the latter case, let $S' = S \backslash \{u\}$. Then f' and S' satisfy the condition of Lemma 3.10.7. Since $\sum_{v \in S'} f'(v) < \sum_{v \in S} f(v)$, by induction hypothesis, there exist $R \subseteq S'$ and g for which the following hold:

1. $g \leq f'$ and $g(v) < f'(v)$ for any $v \in S' \backslash R$.

2. $\mathrm{diff}_G(g^R) \neq 0$.

3. $g(v) = 0$ for every $v \in R$.

As $g(u) < f(u)$ and $f' \leq f$, this completes the proof of Lemma 3.10.7. ∎

By Lemma 3.10.5, $g(v) = 0$ for every $v \in R$ and $\mathrm{diff}_G(g^R) \neq 0$ imply that R is an independent set of G.

Combining Lemmas 3.10.4 and 3.10.7, we have the following corollary.

Corollary 3.10.8 *Assume that* $f \in \mathbb{N}_0^G$, $S \subseteq V(G)$, *and* $\mathrm{diff}_G(f^S) \neq 0$. *Then there exists an independent set* $R \subseteq S$, $g \in \mathbb{N}_0^G$ *for which the following hold:*

1. $g(v) \leq f(v)$ *for any* $v \in V(G) \backslash S$ *and* $g(v) < f(v)$ *for any* $v \in S \backslash R$.

2. $\mathrm{diff}_G(g) \neq 0$.

Lemma 3.10.9 *Assume that* G' *is obtained from* G *by deleting an edge* $e = uv$. *If* $\mathrm{diff}_G(g) \neq 0$, *then* $\mathrm{diff}_{G'}(\hat{g}_u) \neq 0$ *or* $\mathrm{diff}_{G'}(\hat{g}_v) \neq 0$.

Proof. We may assume that $u < v$. For orientation $D \in \mathcal{O}_G(g)$, let $\phi(D)$ be the orientation of G' obtained from D by deleting the edge uv. If e is oriented as (u,v) in D, then $\phi(D) \in \mathcal{O}_{G'}(\hat{g}_u)$ and hence D and $\phi(D)$ have the same parity. If e is oriented as (v,u) in D, then $\phi(D) \in \mathcal{O}_{G'}(\hat{g}_v)$ and $D, \phi(D)$ have opposite parities. Thus ϕ is a one-to-one correspondence from $\mathcal{O}_{e,G}(g)$ to $\mathcal{O}_{e,G'}(\hat{g}_u) \cup \mathcal{O}_{o,G'}(\hat{g}_v)$, and from $\mathcal{O}_{o,G}(g)$ to $\mathcal{O}_{o,G'}(\hat{g}_u) \cup \mathcal{O}_{e,G'}(\hat{g}_v)$. Note that the unions above are disjoint unions. So if $\mathrm{diff}_{G'}(\hat{g}_u) = 0$ and $\mathrm{diff}_{G'}(\hat{g}_v) = 0$, we would have $\mathrm{diff}_G(g) = 0$. ∎

Corollary 3.10.10 *If* G' *is obtained from* G *by deleting a set* F *of edges, and* $\mathrm{diff}_G(g) \neq 0$, *then there exists* $h \in N_0^{G-F}$ *such that* $h \leq g$ *and* $\mathrm{diff}_{G-F}(h) \neq 0$.

Corollary 3.10.11 *If* R *is an independent set and* $\mathrm{diff}_G(g) \neq 0$, *then there exists* $q \in N_0^{G-R}$ *such that* $q(v) \leq g(v)$ *for* $v \in V(G) \backslash R$ *and* $\mathrm{diff}_{G-R}(q) \neq 0$.

Proof. Apply Corollary 3.10.10 to the set F of edges connecting vertices of R to $V(G) \backslash R$ to obtain a function $h \in N_0^{G-F}$ such that $h \leq g$ and $\mathrm{diff}_{G-F}(h) \neq 0$. Since the vertices of R are isolated vertices in $G - F$, we have $\mathrm{diff}_{G-R}(q) \neq 0$, where q is the restriction of h to $V(G - R)$. ∎

Now we are ready to prove the main theorem.

Theorem 3.10.12 *If* G *is* f-AT, *then* G *is* f-paintable.

Proof. We prove this theorem by induction on the number of vertices of G. If G has one vertex, then there is nothing to be proved. Assume that $|V(G)| = n$ and the theorem holds for graphs with fewer vertices.

Assume that G is f-AT. Then there is an orientation D of G such that $d_D^+(v) \leq f(v)$ for each vertex v and $|\mathcal{E}_e(D)| \neq |\mathcal{E}_o(D)|$. Let $g(v) = d_D^+(v)$ for every vertex v. Then for any $S \subseteq V(G)$, any $D' \in \mathcal{O}_G(g^S)$ has the same out-degree sequence as D, and hence is obtained from D by reversing the directions of arcs in an Eulerian subdigraph H of D. If H is an even Eulerian subdigraph of D, then D' has the same parity as D while if H is an odd Eulerian subdigraph of D, then D' and D have opposite parities. So $\operatorname{diff}_G(g^S) = \pm(|\mathcal{E}_e(D)| - |\mathcal{E}_o(D)|)$. Hence $\operatorname{diff}_G(g^S) \neq 0$.

Assume that in the first move, the subset of uncoloured vertices chosen by Lister is S. Since $\operatorname{diff}_G(g^S) \neq 0$, by Corollary 3.10.8 and Corollary 3.10.11, there is an independent set R of G contained in S and a function $q \in N_0^{G-R}$ such that $\operatorname{diff}_{G-R}(q) \neq 0$, and $q(v) \leq g(v)$ for all $v \in V(G-R)$ and $q(v) < g(v)$ for $v \in S \backslash R$. Note that $\operatorname{diff}_{G-R}(q) \neq 0$ implies that there is an orientation D' of $G - R$ such that $d_{D'}^+(v) = q(v)$ for each vertex v and $\operatorname{diff}_{G-R}(q) = \pm(|\mathcal{E}_e(D')| - |\mathcal{E}_o(D')|) \neq 0$. By induction hypothesis, Painter has a winning strategy for the q-painting game on $G \backslash R$.

After the first round, each vertex v of $S \backslash R$ has $f(v) - 1 \geq q(v)$ tokens, and every other vertex v has $f(v) \geq q(v)$ tokens. So by following Painter's strategy for the q-painting game on $G - R$, Painter will win this game. ∎

The concept of d-defective k-choosability is naturally extended to d-defective k-paintability. A *d-defective k-painting game* on a graph G is played by two players: Lister and Painter. The rule is basically the same as the k-painting game on G, except that in each round, after Lister has chosen a subset M, the subset I of M of vertices coloured by Painter is not necessarily an independent set. Instead, it is required that the subgraph $G[I]$ of G induced by I has maximum degree at most d. We say that G is *d-defective k-paintable* if Painter has a winning strategy in the d-defective k-painting game on G. So 0-defective k-painting game is the same as the k-painting game and 0-defective k-paintable is the same as k-paintable.

By the same argument as in the proof of Lemma 3.10.3, we can show that d-defective k-paintability implies d-defective k-choosability. The converse is not true. It is proved in [27] that there are planar graphs that are not 2-defective 3-paintable. Also it is proved in [58] and [21] that every planar graph is 2-defective 3-choosable. Nevertheless, the following lemma holds.

Lemma 3.10.13 *If G has a subgraph H with maximum degree d and $G - E(H)$ is k-paintable, then G is d-defective k-paintable. In particular, if $G - E(H)$ has Alon–Tarsi number at most k, then G is d-defective k-paintable.*

Proof. Painter's winning strategy for the k-painting game on $G - E(H)$ is a winning strategy for the d-defective k-painting game on G. ∎

Corollary 3.10.14 *Every planar graph is 1-defective 4-paintable.*

Chapter 4

Generalizations of CNS and Applications

This chapter introduces some strenghtenings and generalizations of CNS and their applications.

4.1 Number of non-zero points

Assume that $\mathbb{F}$ is a field and $f(\mathbf{x}) = f(x_1, x_2, \ldots, x_n)$ is a polynomial in $\mathbb{F}[x_1, x_2, \ldots, x_n]$. Assume that $\mathbf{S} = S_1 \times S_2 \times \ldots \times S_n$ is a finite grid in $\mathbb{F}^n$. It follows from CNS that if $\mathbf{x}^{\|\mathbf{S}\|}$ is a highest degree non-vanishing monomial in the expansion of $f(\mathbf{x})$, then $f(\mathbf{x})$ has a non-zero point in $\mathbf{S}$. In many cases, $f(\mathbf{x})$ has many non-zero points in $\mathbf{S}$. For example, suppose that G is a planar graph and L is a 5-list assignment of G. By Thomassen's Theorem, if we fix any two adjacent vertices u, v of G, then for any $c \in L(u), c' \in L(v)$ with $c \neq c'$, G has an L-colouring ϕ for which $\phi(u) = c$ and $\phi(v) = c'$. As there are 5 choices of c and at least 4 choices of c', we conclude that G has at least 20 L-colourings. Intuitively, the number of L-colourings of G should be much more than 20, especially when G has many vertices. In this section, we obtain a lower bound on the number of non-zero points of a polynomial in a finite grid $\mathbf{S}$. This result is proved in [13] and a generalization of this result can be found in [12]. The proof presented below is based on a note of Bosek, Grytczuk, Gutowski and Serra [13].

Theorem 4.1.1 *Assume that $\mathbb{F}$ is a field and $f(\mathbf{x}) = f(x_1, x_2, \ldots, x_n)$ is a polynomial in $\mathbb{F}[x_1, x_2, \ldots, x_n]$. Assume further that $S_1, S_2, \ldots, S_n$ are subsets of $\mathbb{F}$ and $|S_i| = s_i$ for each i. Let $d = \deg f$ be the degree of f, $t = \max_{i=1}^n s_i$ and $s = \sum_{i=1}^n s_i$.*

If $\mathbf{S} := S_1 \times S_2 \times \ldots \times S_n$ *contains a non-zero point of* f, *then* $\mathbf{S}$ *contains at least* $t^{(s-n-d)/(t-1)}$ *non-zero points of* f.

Proof. Assume that the theorem is not true and that f and $\mathbf{S}$ form a counterexample with (n, d, s) minimal in lexicographic order. It is obvious that $(n, d, s) > (1, 1, 1)$.

Without loss of generality, assume that $|S_1| \geq 2$. Fix $a \in S_1$ and write

$$f(\mathbf{x}) = (x_1 - a)g(\mathbf{x}) + R(\mathbf{x}), \tag{4.1}$$

Note that x_1 is absent in $R(\mathbf{x})$.

First we consider the case when for any $\mathbf{x} \in S_2 \times S_3 \times \ldots \times S_n$, $R(\mathbf{x}) = 0$. Then for $\mathbf{x} \in (S_1 - \{a\}) \times S_2 \times \ldots \times S_n$, $f(\mathbf{x}) \neq 0$ if and only if $g(\mathbf{x}) \neq 0$. For $g(\mathbf{x})$ and $S_1 - \{a\}, S_2, \ldots, S_n$, the corresponding parameters s', n', d', t' satisfy the following: $s' = s - 1, n' = n, d' = d - 1, t' \leq t$. By induction hypothesis, $(S_1 - \{a\}) \times S_2 \times \ldots \times S_n$ contains at least $(t')^{(s'-n'-d')/(t'-1)}$ non-zero points of $g(\mathbf{x})$. Observe that $t^{1/(t-1)}$ is a decreasing function. Hence $(t')^{1/(t'-1)} \geq t^{1/(t-1)}$ and $(t')^{(s'-n'-d')/(t'-1)} \geq t^{(s-n-d)/(t-1)}$, and we are done.

Assume now that $R(\mathbf{x})$ has a non-zero point in $S_2 \times S_3 \times \ldots \times S_n$. The corresponding parameters s', n', d', t' for $R(\mathbf{x})$ satisfy the following: $s' = s - s_1, n' = n - 1, d' \leq d, t' \leq t$. By induction hypothesis, $R(\mathbf{x})$ has at least $(t')^{(s'-n'-d')/(t'-1)}$ non-zero points in $S_2 \times S_3 \times \ldots \times S_n$. Since $(t')^{1/(t'-1)} \geq t^{1/(t-1)}$, we have $(t')^{(s'-n'-d')/(t'-1)} \geq t^{\{(s-n-d)/(t-1)\}-\{(s_1-1)/(t-1)\}}$. If $(a_2, a_3, \ldots, a_n) \in S_2 \times S_3 \times \ldots \times S_n$ is a non-zero point for $R(\mathbf{x})$, then $(a, a_2, a_3, \ldots, a_n)$ is a non-zero point for $f(\mathbf{x})$. So $f(\mathbf{x})$ has at least $t^{\{(s-n-d)/(t-1)\}-\{(s_1-1)/(t-1)\}}$ non-zero points in $\{a\} \times S_2 \times S_3 \times \ldots \times S_n$.

We apply the same argument for each element a of S_1. Then either for some $a \in S_1$, the corresponding $R(\mathbf{x})$ has no non-zero point in $S_2 \times S_3 \times \ldots \times S_n$, and we are done. Otherwise, $f(\mathbf{x})$ has at least $t^{\{(s-n-d)/(t-1)\}-\{s_1/(t-1)\}}$ non-zero points in $\{a\} \times S_2 \times S_3 \times \ldots \times S_n$ for each $a \in S_1$. Since $s_1 \leq t$, we have $t^{1/(t-1)} \leq s_1^{1/(s_1-1)}$. Hence $t^{(s_1-1)/(t-1)} \leq s_1$ and $f(\mathbf{x})$ has at least $s_1 \times (t^{\{(s-n-d)/(t-1)\}-\{(s_1-1)/(t-1)\}}) \geq t^{(s-n-d)/(t-1)}$ non-zero points in $S_1 \times S_2 \times \ldots \times S_n$. ∎

Corollary 4.1.2 *Assume that* G *is an* n-*vertex planar graph and* L *is a* 5-*list assignment of* G. *Then* G *has at least* $5^{n/4}$ L-*colourings.*

Proof. Apply Theorem 4.1.1 to the graph polynomial f_G of G. Since $s = 5n, d < 3n, t = 5$, we conclude that G has at least $5^{n/4}$ L-colourings. ∎

4.2 Multisets

Consider the polynomial $f(x) = a_k x^k + a_{k-1} x^{k-1} + \ldots + a_1 x + a_0 \in \mathbb{F}[x]$ of degree k in the single variable x. Then f has at most k distinct roots in the field $\mathbb{F}$. Thus if S is a subset of $\mathbb{F}$ of cardinality $k + 1$, then S contains a non-zero point of $f(x)$. The CNS is basically an extension of this result to polynomials with n variables. Now if $f(x)$ has repeated roots, then the number of distinct roots of $f(x)$ is smaller than k. Hence if S is a subset of $\mathbb{F}$ of cardinality k, then S contains a non-zero point of $f(x)$. This section discusses extension of this result to polynomials with more variables.

Let $\mathbb{F}$ be a field. A multi-subset of $\mathbb{F}$ is a mapping $\omega : \mathbb{F} \to \mathbb{N}$, where $\omega(a)$ is the multiplicity of a. The support of ω is $\text{supp}(\omega) = \{a \in \mathbb{F} : \omega(a) \geq 1\}$. The *size* of ω is $|\omega| = \sum_{a \in \mathbb{F}} \omega(a)$.

For $f(\mathbf{x}) \in \mathbb{F}[x_1, x_2, \ldots, x_n]$, for $i = 1, 2, \ldots, n$, $\frac{\partial^0}{\partial x_i^0} f(\mathbf{x}) = f(\mathbf{x})$, and for $k \geq 1$,

$$\frac{\partial^k}{\partial x_i^k} f(\mathbf{x}) = \frac{\partial}{\partial x_i} \left(\frac{\partial^{k-1}}{\partial x_i^{k-1}} f(\mathbf{x}) \right).$$

For $\mathbf{u} = (u_1, u_2, \ldots, u_n) \in \mathbb{N}_0^n$,

$$\frac{\partial^{\mathbf{u}}}{\partial \mathbf{x}^{\mathbf{u}}} f(\mathbf{x}) = \frac{\partial^{u_1 + u_2 + \ldots + u_n}}{\partial x_1^{u_1} \partial x_2^{u_2} \ldots \partial x_n^{u_n}} f(\mathbf{x}) = \frac{\partial^{u_1}}{\partial x_1^{u_1}} \left(\frac{\partial^{u_2 + \ldots + u_n}}{\partial x_2^{u_2} \ldots \partial x_n^{u_n}} f(\mathbf{x}) \right).$$

Lemma 4.2.1 *Assume that $g(\mathbf{x}) \in \mathbb{F}[x_1, x_2, \ldots, x_n]$ and $u_1 \geq 1$. Then*

$$\frac{\partial^{u_1}}{\partial x_1^{u_1}} \left((x_1 - a) g(\mathbf{x}) \right) = u_1 \frac{\partial^{u_1 - 1}}{\partial x_1^{u_1 - 1}} g(\mathbf{x}) + (x_1 - a) \frac{\partial^{u_1}}{\partial x_1^{u_1}} g(\mathbf{x}).$$

Proof. By induction on u_1. If $u_1 = 1$, then this is just the product rule. Assume that $u_1 \geq 2$ and the claim holds for $u_1 - 1$. Then

$$
\begin{aligned}
\frac{\partial^{u_1}}{\partial x_1^{u_1}} \left((x_1 - a)g(\mathbf{x}) \right) &= \frac{\partial}{\partial x_1} \left((u_1 - 1) \frac{\partial^{u_1 - 2}}{\partial x_1^{u_1 - 2}} g(\mathbf{x}) \right. \\
&\quad \left. + (x_1 - a) \frac{\partial^{u_1 - 1}}{\partial x_1^{u_1 - 1}} g(\mathbf{x}) \right) \\
&= (u_1 - 1) \frac{\partial^{u_1 - 1}}{\partial x_1^{u_1 - 1}} g(\mathbf{x}) + (x_1 - a) \frac{\partial^{u_1}}{\partial x_1^{u_1}} g(\mathbf{x}) \\
&\quad + \frac{\partial^{u_1 - 1}}{\partial x_1^{u_1 - 1}} g(\mathbf{x}) \\
&= u_1 \frac{\partial^{u_1 - 1}}{\partial x_1^{u_1 - 1}} g(\mathbf{x}) + (x_1 - a) \frac{\partial^{u_1}}{\partial x_1^{u_1}} g(\mathbf{x}).
\end{aligned}
$$

∎

The following result was proved by Kós and Rónyai [40]. The proof given below is new and it is adapted from the proof of Theorem 2.1.3 of Michałek [47].

Theorem 4.2.2 *[Combinatorial Nullstellensatz for multisets (CNSM)] Let $\mathbb{F}$ be a field and let $f(\mathbf{x}) = f(x_1, x_2, \ldots, x_n)$ be a nonzero polynomial in $\mathbb{F}[x_1, x_2, \ldots, x_n]$. Suppose that $x_1^{t_1} x_2^{t_2} \ldots x_n^{t_n}$ is a highest degree non-vanishing monomial of f. Then for any multisubsets $\omega_1, \omega_2, \ldots, \omega_n$ of $\mathbb{F}$ with $|\omega_i| = t_i + 1$ for each i, there exist $\mathbf{s} \in \mathbb{F}^n$ and $\mathbf{u} \in \mathbb{N}_0^n$ with $u_i < \omega_i(s_i)$ for $i = 1, 2, \ldots, n$ such that*

$$
\frac{\partial^{\mathbf{u}}}{\partial \mathbf{x}^{\mathbf{u}}} f(\mathbf{s}) \neq 0.
$$

Proof. Observe that for each i, $0 \leq u_i < \omega_i(s_i)$. So $\omega_i(s_i) \geq 1$, i.e., $s_i \in \text{supp}(\omega_i)$.

The proof is by induction on the degree of f.

Assume that $\deg f = 1$ so that

$$
f(\mathbf{x}) = a_1 x_1 + a_2 x_2 + \cdots + a_n x_n + a_{n+1},
$$

$a_i \in \mathbb{F}$ with at least one $a_i \in \{a_1, \ldots, a_n\}$, say $a_1 \neq 0$. Assume that $\omega_1, \omega_2, \ldots, \omega_n$ are multi-subsets of $\mathbb{F}$ with $|\omega_i| \geq 1$ for $i = 2, 3, \ldots, n$ and $|\omega_1| \geq 2$. For $i = 2, 3, \ldots, n$, assign any value s_i to x_i for which $\omega_i(s_i) \geq 1$. This gives the element $a_2 s_2 + \cdots + a_n s_n + a_{n+1}$ of $\mathbb{F}$. If $|\text{supp}(\omega_1)| = 2$, then there exists $s_1 \in \text{supp}(\omega_1)$ so that

$$
a_1 s_1 + a_2 s_2 + \cdots + a_n s_n + a_{n+1} \neq 0.
$$

Otherwise, $\omega(s_1) = 2$ for some $s_1 \in \mathbb{F}$, and

$$\frac{\partial}{\partial x_1} f(s_1, s_2, \ldots, s_n) = a_1 \neq 0.$$

Hence the result is true when $\deg f = 1$.

Now assume that $f(\mathbf{x}) \in \mathbb{F}[x_1, x_2, x_3, \ldots, x_n]$ is a polynomial of degree $\deg f = r$ for some $r \geq 2$, and Theorem 4.2.2 holds for polynomials of degree $\leq r - 1$. Suppose that $x_1^{t_1} x_2^{t_2} \ldots x_n^{t_n}$ is a highest degree non-vanishing monomial of f. Without loss of generality, assume that $t_1 > 0$.

Suppose that f does not satisfy the statement in the theorem and there exist multi-subsets $\omega_1, \omega_2, \ldots, \omega_n$ of $\mathbb{F}$ with $|\omega_i| = t_i + 1$ such that for any $\mathbf{u} \in \mathbb{N}_0^n$ and any $\mathbf{s} \in \mathbb{F}^n$ for which $u_i < \omega_i(s_i)$ for $i = 1, 2, \ldots, n$,

$$\frac{\partial^{\mathbf{u}}}{\partial \mathbf{x}^{\mathbf{u}}} f(\mathbf{s}) = 0. \tag{4.2}$$

Fix $a \in \text{supp}(\omega_1)$ and write

$$f(\mathbf{x}) = (x_1 - a)g(\mathbf{x}) + R(\mathbf{x}), \tag{4.3}$$

As $t_1 > 0$, g has a non-vanishing monomial $x_1^{t_1-1} x_2^{t_2} \ldots x_n^{t_n}$, and

$$\begin{aligned} \deg g &= (t_1 - 1) + t_2 + \cdots + t_n \\ &= \deg f - 1. \end{aligned}$$

Let ω_1' be the multi-subset defined as $\omega_1'(a) = 1$ and $\omega_1'(s) = 0$ for $s \in \mathbb{F} - \{a\}$, and for $i = 2, 3, \ldots, n$, $\omega_i' = \omega_i$.

Assume that $\mathbf{u}' \in \mathbb{N}_0^n$, $\mathbf{s} \in \mathbb{F}^n$, and $u_i' < \omega_i'(s_i)$ for $i = 1, 2, \ldots, n$. Since $\omega_1'(a) = 1$ and $\omega_1'(s) = 0$ for $s \in \mathbb{F} - \{a\}$, we have $u_1' = 0$ and $s_1 = a$. Hence

$$\frac{\partial^{\mathbf{u}'}}{\partial \mathbf{x}^{\mathbf{u}'}} ((x_1 - a)g(\mathbf{x}))|_{\mathbf{x}=\mathbf{s}} = 0.$$

By assumption,

$$\frac{\partial^{\mathbf{u}'}}{\partial \mathbf{x}^{\mathbf{u}'}} f(\mathbf{s}) = 0.$$

Therefore

$$\frac{\partial^{\mathbf{u}'}}{\partial \mathbf{x}^{\mathbf{u}'}} R(\mathbf{s}) = 0.$$

As $R(\mathbf{x})$ does not contain x_1, it follows that for any $\mathbf{s} \in \mathbb{F}^n$ and $\mathbf{u} \in \mathbb{N}_0^n$ with $u_i < \omega_i(s_i)$,

$$\frac{\partial^{\mathbf{u}}}{\partial \mathbf{x}^{\mathbf{u}}} R(\mathbf{s}) = 0.$$

Plugging this and (4.2) into (4.3), we have

$$\frac{\partial^{\mathbf{u}}}{\partial \mathbf{x}^{\mathbf{u}}} ((x_1 - a)g(\mathbf{x}))|_{\mathbf{x}=\mathbf{s}} = 0. \tag{4.4}$$

Let $\omega_1''(a) = \omega_1(a) - 1$ and for $b \in \mathbb{F} - \{a\}$, $\omega_1''(b) = \omega_1(b)$. So $|\omega_1''| = |\omega_1| - 1 > t_1 - 1$. By induction hypothesis, there exist $\mathbf{u}'' \in \mathbb{N}_0^n$ and $\mathbf{s}' \in \mathbb{F}^n$ such that $u_1'' < \omega_1''(s_1')$ and $u_i'' < \omega_i(s_i')$ and

$$\frac{\partial^{\mathbf{u}''}}{\partial \mathbf{x}^{\mathbf{u}''}} g(\mathbf{s}') \neq 0. \tag{4.5}$$

We choose a minimal $\mathbf{u}''$ so that (4.5) holds, i.e., for any $\mathbf{u}^* < \mathbf{u}''$,

$$\frac{\partial^{\mathbf{u}^*}}{\partial \mathbf{x}^{\mathbf{u}^*}} g(\mathbf{s}') = 0.$$

If $s_1' = a$, then let $\mathbf{u} = (u_1'' + 1, u_2'', \dots, u_n'')$. Otherwise let $\mathbf{u} = \mathbf{u}''$. Then $u_i < \omega_i(s_i')$ for $i = 1, 2, \dots, n$.

If $s_1' = a$, then $u_1 = u_1'' + 1$ and by Lemma 4.2.1,

$$\frac{\partial^{\mathbf{u}}}{\partial \mathbf{x}^{\mathbf{u}}} ((x_1 - a)g(\mathbf{x}))|_{\mathbf{x}=\mathbf{s}'} = u_1 \frac{\partial^{\mathbf{u}''}}{\partial \mathbf{x}^{\mathbf{u}''}} g(\mathbf{s}') \neq 0,$$

contrary to (4.4).

Assume that $s_1' \neq a$. Thus $\mathbf{u} = \mathbf{u}''$. If $u_1 = 0$, then

$$\left(\frac{\partial^{\mathbf{u}}}{\partial \mathbf{x}^{\mathbf{u}}} ((x_1 - a)g(\mathbf{s}))\right)|_{\mathbf{x}=\mathbf{s}'} = \left((x_1 - a)\frac{\partial^{\mathbf{u}}}{\mathbf{x}^{\mathbf{u}}} g(\mathbf{s})\right)|_{\mathbf{x}=\mathbf{s}'} \neq 0,$$

contrary to (4.4).

Assume that $u_1 = u_1'' \geq 1$. Let $u_1^* = u_1 - 1$ and $u_i^* = u_i$ for $i = 2, 3, \dots, n$. Then by Lemma 4.2.1,

$$\frac{\partial^{\mathbf{u}}}{\partial \mathbf{x}^{\mathbf{u}}} ((x_1 - a)g(\mathbf{s})) = u_1 \frac{\partial^{\mathbf{u}^*}}{\partial \mathbf{x}^{\mathbf{u}^*}} g(\mathbf{s}') + (s_1' - a)\frac{\partial^{\mathbf{u}}}{\partial \mathbf{x}^{\mathbf{u}}} g(\mathbf{s}').$$

Note that since $\mathbf{u}^* < \mathbf{u}''$, by our choice of $\mathbf{u}''$,

$$\frac{\partial^{\mathbf{u}^*}}{\partial \mathbf{x}^{\mathbf{u}^*}} g(\mathbf{s}') = 0.$$

Hence

$$\frac{\partial^{\mathbf{u}}}{\partial \mathbf{x}^{\mathbf{u}}}\left((x_1 - a)g(\mathbf{s})\right) = (s_1' - a)\frac{\partial^{\mathbf{u}}}{\partial \mathbf{x}^{\mathbf{u}}}g(\mathbf{s}') \neq 0,$$

contrary to (4.4). This contradiction proves the theorem. ∎

Some further generalizations of CNS to multisets are discussed in [11, 39].

4.3 Coefficient of a highest degree monomial

The coefficient of a highest degree monomial in the expansion of a polynomial $f(x_1, x_2, \ldots, x_n)$, which is crucial when we invoke CNS, can be expressed by means of an interpolation of f at appropriate sets of points. This chapter explores applications of such interpolation formulas.

Assume that $f(\mathbf{x}) \in \mathbb{F}[x_1, x_2, \ldots, x_n]$ is a polynomial over $\mathbb{F}$, with degree $\deg f \leq d_1 + d_2 + \ldots + d_n$. Uwe Schauz [55] and independently Michał Lasoń [44] proved a strengthening of CNS, which shows that the coefficient of the monomial $\prod_{i=1}^{n} x_i^{d_i}$ in the expansion of f can be written as an interpolation of its values on any grid of size $(d_1, d_2, \ldots, d_n)$.

Let $\mathbf{S} := S_1 \times S_2 \times \cdots \times S_n$, where each S_i is a subset of $\mathbb{F}$. For $\mathbf{a} := (a_1, a_2, \ldots, a_n) \in \mathbf{S}$, let

$$N_{\mathbf{S}}(\mathbf{a}) = \prod_{i=1}^{n} \prod_{b \in S_i - \{a_i\}} (a_i - b).$$

We may think of $N_{\mathbf{S}}(\mathbf{a})$ as a normalizing factor w.r.t. $\mathbf{a}$. Let

$$\chi_{\mathbf{a}}(\mathbf{x}) = N_{\mathbf{S}}(\mathbf{a})^{-1} \prod_{i=1}^{n} \prod_{b \in S_i - \{a_i\}} (x_i - b).$$

Then $\chi_{\mathbf{a}}$ vanishes at all points of the grid $\mathbf{S}$, except that $\chi_{\mathbf{a}}(\mathbf{a}) = 1$.

Theorem 4.3.1 *Assume that $f(\mathbf{x}) \in \mathbb{F}[x_1, x_2, \ldots, x_n]$ is a polynomial over $\mathbb{F}$, $\mathbf{S} := S_1 \times S_2 \times \cdots \times S_n$ is an n-dimensional grid over $\mathbb{F}$, and $\deg f \leq s = \sum_{i=1}^{n} |S_i|$. Then the coefficient $c_{f,\|S\|}$ of the monomial $\prod_{i=1}^{n} x_i^{|S_i|-1}$ in the expansion of $f(\mathbf{x})$ is given by*

$$c_{f,\|\mathbf{S}\|} = \sum_{\mathbf{a} \in \mathbf{S}} N_{\mathbf{S}}(\mathbf{a})^{-1} f(\mathbf{a}).$$

Proof. For $i \in \{1, 2, \ldots, n\}$, let $d_i = |S_i| - 1$. Assume first that for each $i \in \{1, 2, \ldots, n\}$, the highest degree of x_i in f is at most d_i. Let

$$p(\mathbf{x}) = \sum_{\mathbf{a} \in \mathbf{S}} \chi_{\mathbf{a}}(\mathbf{x}) f(\mathbf{a}).$$

By definition, $p(\mathbf{x})$ is a polynomial in which the highest degree of x_i is at most d_i. Since $\chi_{\mathbf{a}}(\mathbf{x}) = 1$ if $\mathbf{x} = \mathbf{a}$ and $\chi_{\mathbf{a}}(\mathbf{x}) = 0$ if $\mathbf{x} \in \mathbf{S} - \{\mathbf{a}\}$, we have $p(\mathbf{a}) = f(\mathbf{a})$ for $\mathbf{a} \in \mathbf{S}$. Thus $f - p$ is a polynomial in which the highest degree of x_i is at most d_i, and $(f - p)(\mathbf{a}) = 0$ for all $\mathbf{a} \in \mathbf{S}$. If there is only one variable, then $f - p$ is a polynomial of degree d_1 in x_1 and with $d_1 + 1$ roots. So $f - p = 0$ is the zero polynomial. If there are more than one variable, then it is easy to prove by induction on the number of variables that $f - p = 0$ is the zero polynomial, i.e., $f = p$. (We can also apply Theorem 2.1.3 to conclude that $f - p = 0$ is the zero polynomial.) The coefficient of $\prod_{i=1}^{n} x_i^{d_i}$ in $p(\mathbf{x})$ is easily seen to be $\sum_{\mathbf{a} \in \mathbf{S}} N_{\mathbf{S}}(\mathbf{a})^{-1} f(\mathbf{a})$. So the theorem holds if the highest degree of x_i in f is at most d_i for each $i \in \{1, 2, \ldots, n\}$.

So assume now that there is an index $j \in \{1, 2, \ldots, n\}$ such that f has a non-vanishing monomial $\prod_{i=1}^{n} x_i^{s_i}$ with $s_j \geq d_j + 1$. Let

$$g_j(x_j) = \prod_{a \in S_j} (x_j - a), \text{ and } h_j(x_j) = x_j^{d_j + 1} - g_j(x_j).$$

Then h_j is a polynomial of degree d_j. For any $a \in S_j$, since $g_j(a) = 0$, we have $a^{d_j + 1} = h_j(a)$.

In the expansion of f, for each non-vanishing monomial $\prod_{i=1}^{n} x_i^{s_i}$ for which $s_j > d_j$, we replace $x_j^{s_j}$ by $x_j^{s_j - d_j - 1} h_j(x_j)$. Denote the resulting polynomial by $\tilde{f}$.

Then the following hold:

(1) For any $\mathbf{a} \in \mathbf{S}$, $f(\mathbf{a}) = \tilde{f}(\mathbf{a})$.

(2) The coefficients of $\prod_{i=1}^{n} x_i^{d_i}$ in the expansions of f and $\tilde{f}$ are the same.

(1) holds because for $a \in S_j$, $a^{d_j + 1} = h_j(a)$. So replacing $x_j^{d_j + 1}$ by $h_j(x_j)$ does not change the value of $f(\mathbf{a})$ for $\mathbf{a} \in \mathbf{S}$.

(2) holds because each monomial $\prod_{i=1}^{n} x_i^{s_i}$ that we have changed is different from the monomial $\prod_{i=1}^{n} x_i^{d_i}$, and the polynomial that we used to replace $\prod_{i=1}^{n} x_i^{s_i}$ is not of highest degree in f. So this change does not affect the coefficient of $\prod_{i=1}^{n} x_i^{d_i}$.

Repeat this process whenever there is an index j for which the current polynomial $\tilde{f}$ has a non-vanishing monomial $\prod_{i=1}^{n} x_i^{s_i}$ for which $s_j > d_j$. Eventually, we arrive at a polynomial $\tilde{f}$ in which x_j has degree at most d_j, and the following hold:

- For any $\mathbf{a} \in \mathbf{S}$, $f(\mathbf{a}) = \tilde{f}(\mathbf{a})$.

- The coefficient of $\prod_{i=1}^{n} x_i^{d_i}$ in the expansions of f and $\tilde{f}$ are the same.

By the first half of this proof, the coefficient of the monomial $\prod_{i=1}^{n} x_i^{d_i}$ in the expansion of $\tilde{f}$ is

$$c_{\tilde{f}, \|\mathbf{s}\|} = \sum_{\mathbf{a} \in \mathbf{S}} N_{\mathbf{S}}(\mathbf{a})^{-1} \tilde{f}(\mathbf{a}) = \sum_{\mathbf{a} \in \mathbf{S}} N_{\mathbf{S}}(\mathbf{a})^{-1} f(\mathbf{a}) = c_{f, \|\mathbf{s}\|}.$$

As the coefficients of $\prod_{i=1}^{n} x_i^{d_i}$ in the expansion of f and $\tilde{f}$ are the same, we are done.   ∎

Note that Theorem 2.1.3 (CNS) is a corollary of Theorem 4.3.1. Assume that $f(\mathbf{x}) \in \mathbb{F}[x_1, x_2, \ldots, x_n]$ is a polynomial over $\mathbb{F}$, and $d_i \geq 0$ are integers such that $\deg f \leq \sum_{i=1}^{n} d_i$. Suppose that $\mathbf{S} := S_1 \times S_2 \times \cdots \times S_n$ is an n-dimensional grid over $\mathbb{F}$ with $|S_i| = d_i + 1$. If $f(\mathbf{a}) = 0$ for all $\mathbf{a} \in \mathbf{S}$, then it follows from Theorem 4.3.1 that the coefficient of $\prod_{i=1}^{n} x_i^{d_i}$ is 0.

Also, it follows from Theorem 4.3.1 that if $c_{f, \|\mathbf{S}\|} = 0$, then either $\mathbf{S}$ contains no non-zero point of f or contains at least two non-zero points of f.

4.4 Calculation of $N_{\mathbf{S}}(\mathbf{a})$

To apply Theorem 4.3.1, we need to calculate $N_{\mathbf{S}}(\mathbf{a})$ for all $\mathbf{a} \in \mathbf{S}$. This seems to be a nontrivial task. However, in some cases, we have simple formulas for $N_{\mathbf{S}}(\mathbf{a})$. Lemma 4.4.1 below, proved in [55], shows that for some special sets S, there is a simple formula for $\prod_{b \in S - \{a\}} (a - b)$.

For subset S of $\mathbb{F}$, let $f_S(x) = \prod_{b \in S} (x - b)$. Then $N_{\mathbf{S}}(\mathbf{a}) = \prod_{i=1}^{n} f_{S_i - \{a_i\}}(a_i)$. The following lemma gives formulas for $f_{S - \{a\}}(a)$ for some special sets S.

Lemma 4.4.1 *Assume that $\mathbb{F}$ is a field. For a positive integer k, let $R_k = \{c \in \mathbb{F} : c^k = 1\}$ be the set of kth roots of unity in the field $\mathbb{F}$. Assume that $|R_k| = k$.*

1. *If $S = R_k$, then for $a \in S$, $f_{S-\{a\}}(a) = ka^{-1}$. In particular, if $R_k \cup \{0\}$ is a finite subfield of $\mathbb{F}$, then $f_{S-\{a\}}(a) = -a^{-1}$.*

2. *If $S = R_k \cup \{0\}$, then for $a \in S - \{0\}$, $f_{S-\{a\}}(a) = k1$, and $f_{S-\{0\}}(0) = -1$.*

3. *If S is a subfield of $\mathbb{F}$, then for $a \in S$, $f_{S-\{a\}}(a) = -1$.*

4. *If $S = \{0, 1, \ldots, d\}$, then for $a \in S$, $f_{S-\{a\}}(a) = (-1)^{d+a} d! \binom{d}{a}^{-1}$.*

Proof. (1) Since

$$f_{R_k}(x) = \prod_{b \in R_k} (x - b) = x^k - 1 = (x - 1)(x^{k-1} + x^{k-2} + \ldots + x^0),$$

we conclude that $f_{R_k - \{1\}}(x) = f_{R_k}(x)/(x - 1) = x^{k-1} + x^{k-2} + \ldots + x^0$ and $f_{R_k - \{1\}}(1) = k$.

For $a \in R_k$, $aR_k = \{ba : b \in R_k\} = R_k$ and $a(R_k - \{1\}) = R_k - \{a\}$. Hence $f_{R_k - \{a\}}(a) = f_{a(R_k - \{1\})}(a) = \prod_{b \in R_k - \{1\}} (a - ab) = a^{k-1} f_{R_k - \{1\}}(1) = ka^{k-1} = ka^{-1}$. If $R_k \cup \{0\}$ is a finite subfield $\mathbb{F}_{p^t}$ of $\mathbb{F}$, then $|R_k| = k = p^t - 1 = -1 \pmod{p^t}$, and hence $f_{R_k - \{a\}}(a) = -a^{-1}$. This proves (1).

(2) If $S = R_k \cup \{0\}$, then for $a \in S - \{0\}$, $f_{S-\{a\}}(a) = af_{S_k - \{a\}}(a)$. By (1) $f_{S_k - \{a\}}(a) = ka^{-1}$. Hence $f_{S-\{a\}}(a) = k$. As $f_{S-\{0\}}(x) = f_{R_k}(x) = x^k - 1$, we have $f_{S-\{0\}}(0) = -1$.

(3) If S is a finite subfield $\mathbb{F}_{p^t}$ of $\mathbb{F}$, then $S = R_k \cup \{0\}$ for $k = p^t - 1 = -1$ and the conclusion follows from (2).

(4) Plug in the values in the definition of $f_{S-\{a\}}(a)$, we have

$$
\begin{aligned}
f_{S-\{a\}}(a) &= \left(\prod_{0 \le b < a} (a - b) \right) \prod_{a < b \le d} (a - b) = a!(d - a)!(-1)^{d-a} \\
&= (-1)^{d+a} d! \binom{d}{a}^{-1}.
\end{aligned}
$$

∎

Corollary 4.4.2 *Assume that $\mathcal{F}$ is a finite field of order q, $f(\mathbf{x}) \in \mathcal{F}[x_1, x_2, \ldots, x_n]$ and $\deg f \leq n(q-1)$. Let $\mathbf{t} = (q-1, q-1, \ldots, q-1)$. Then*

$$c_{f,\mathbf{t}} = (-1)^n \sum_{\mathbf{a} \in \mathcal{F}^n} f(\mathbf{a}).$$

Proof. By (3) of Lemma 4.4.1, for $\mathbf{S} = \mathcal{F}^n$, we have $N_{\mathbf{S}}(\mathbf{a}) = (-1)^n$ for every $\mathbf{a} \in \mathbf{S}$. ∎

Corollary 4.4.3 *Assume that $f(\mathbf{x}) \in \mathbb{F}[x_1, x_2, \ldots, x_n]$ is a polynomial over $\mathbb{F}$, and $d_i \geq 0$ are integers such that $\deg f \leq \sum_{i=1}^n d_i$. Let $\mathbf{d} = (d_1, d_2, \ldots, d_n)$. Then the coefficient of the monomial $\prod_{i=1}^n x_i^{d_i}$ in the expansion of f is*

$$c_{f,\mathbf{d}} = \left(\prod_{i=1}^n d_i! \right)^{-1} \sum_{a_1=0}^{d_1} \cdots \sum_{a_n=0}^{d_n} (-1)^{d_1+a_1} \binom{d_1}{a_1}$$
$$\cdots (-1)^{d_n+a_n} \binom{d_n}{a_n} f(a_1, \ldots, a_n).$$

In particular, if $d_i = d$ for all i, then the coefficient of the monomial $\prod_{i=1}^n x_i^d$ in the expansion of f is

$$c_{f,\mathbf{d}} = (d!)^{-n} \sum_{\sigma} \left(\prod_{i=1}^n (-1)^{d+\sigma(x_i)} \binom{d}{\sigma(x_i)} \right) f(\sigma),$$

where the summation is over all mappings $\sigma : \{x_1, x_2, \ldots, x_n\} \to \{0, 1, \ldots, d\}$ and $f(\sigma)$ is the evaluation of f at $\sigma(x_i)$ for $i = 1, 2, \ldots, n$.

Proof. For $S_i = \{0, 1, \ldots, d_i\}$, $1 \leq i \leq n$, and $\mathbf{S} = S_1 \times S_2 \times \ldots \times S_n$, it follows from (4) of Lemma 4.4.1 that $N_{\mathbf{S}}(\mathbf{a}) = \prod_{i=1}^n \frac{1}{d_i!} (-1)^{d_i - a_i} \binom{d_i}{a_i}$ for $\mathbf{a} \in \mathbf{S}$. ∎

Corollary 4.4.3 is a classical result, proved before Theorem 4.3.1. Below is a simple direct proof of this corollary given in [57].

An alternative proof of Corollary 4.4.3

Let $S = \sum_{a_1=0}^{s_1} \cdots \sum_{a_n=0}^{s_n} (-1)^{s_1+a_1} \binom{s_1}{a_1} \cdots (-1)^{s_n+a_n} \binom{s_n}{a_n} f(a_1, \ldots, a_n)$. By linearity, it suffices to show that equality holds when $f(x_1, x_2, \ldots, x_n) = \prod_{i=1}^n x_i^{t_i}$ is a monomial. Assume that $f(x_1, x_2, \ldots, x_n) = \prod_{i=1}^n x_i^{t_i}$. Then

$$c_{f,\mathbf{s}} = \begin{cases} 1, & \text{if } \mathbf{s} = \mathbf{t}, \\ 0, & \text{otherwise.} \end{cases}$$

So we need to show that $S = \prod_{i=1}^{n} s_i!$ if $\mathbf{s} = \mathbf{t}$ and $S = 0$ if $\mathbf{s} \neq \mathbf{t}$. Plugging in the formula $f(x_1, x_2, \ldots, x_n) = \prod_{i=1}^{n} x_i^{t_i}$ into the definition of S, we have

$$S = \sum_{a_1=0}^{s_1} \cdots \sum_{a_n=0}^{s_n} (-1)^{s_1+a_1} \binom{s_1}{a_1} \cdots (-1)^{s_n+a_n} \binom{s_n}{a_n} \prod_{i=1}^{n} a_i^{t_i}$$

$$= \prod_{i=1}^{n} \left(\sum_{j=0}^{s_i} (-1)^{s_i+j} \binom{s_i}{j} j^{t_i} \right).$$

By inclusion-exclusion principle, $\sum_{j=0}^{s_i} (-1)^{s_i+j} \binom{s_i}{j} j^{t_i}$ counts the number of onto mappings from a t_i-set to an s_i-set. If $s_i = t_i$ for all i, then $S = \prod_{i=1}^{n} s_i!$. Otherwise, $t_i < s_i$ for some i, and $S = 0$.
∎

4.5   Alon–Tarsi number of $K_{2\star n}$ and cycle powers

In this section, we use Corollary 4.4.3 to calculate the coefficient of some monomials in graph polynomial.

We denote by $K_{2\star n}$ the complete multipartite graph which has n partite sets and with each part of size 2. By applying Hall's Theorem on the existence of a System of Distinct Representatives [7], it is easy to show that $K_{2\star n}$ is n-choosable. Theorem 4.5.1 determines the Alon–Tarsi number of $K_{2\star n}$ and is given in [29].

Theorem 4.5.1 $AT(K_{2\star n}) = n.$

Proof. Let $G = K_{2\star n}$. Let $V_1, V_2, \ldots V_n$ be the partite sets of G, with $V_i = \{u_i, v_i\}$. Since $\deg f_G = 2n(n-1)$, to prove Theorem 4.5.1, it suffices to show that the monomial $\prod_{v \in V(G)} x_v^{n-1}$, which is of degree $2n(n-1)$, is non-vanishing. By Corollary 4.4.3, this is equivalent to showing that

$$\sum_{\sigma} \prod_{v \in V(G)} (-1)^{n-1+\sigma(v)} \binom{n-1}{\sigma(v)} f_G(\sigma) \neq 0$$

where the summation is over all mappings $\sigma : V(G) \to \{0, 1, \ldots, n-1\}$. However, if σ is not a proper colouring of G, then $f_G(\sigma) = 0$. Hence we may restrict the summation to proper colourings σ of G with colour set $\{0, 1, \ldots, n-1\}$. As G is uniquely n-colourable, all the proper colourings σ of G are of the form $\sigma_\pi(u_i) = \sigma_\pi(v_i) = \pi(i)$, where π is a permutation of $\{0, 1, \ldots, n-1\}$. For each $i \in \{0, 1, \ldots, n-1\}$, there are exactly two vertices of colour i. So we conclude that

$$\prod_{v \in V(G)} (-1)^{n-1+\sigma(v)} \binom{n-1}{\sigma(v)} = \prod_{i=0}^{n-1} \binom{n-1}{i}^2.$$

Also if π' is obtained from π by interchanging colours i, j, then $f_G(\sigma_{\pi'})$ is obtained from $f_G(\sigma_\pi)$ by changing $(i-j)^4$ to $(j-i)^4$. Hence $f_G(\sigma)$ is a non-zero constant (it does not depend on σ). Therefore $\sum_\sigma \prod_{v \in V(G)} (-1)^{n-1+\sigma(v)} \binom{n-1}{\sigma(v)} f_G(\sigma) \neq 0$. ∎

Given a graph G and a positive integer k, the power G^k is the graph with vertex set $V(G)$ in which two distinct vertices u and v are adjacent if and only if $d_G(u, v) \leq k$. Disproving a conjecture of Kostochka and Woodall [42], it was proved by Kim and Park [38, 37] that there are graphs G for which $\chi(G^2) < ch(G^2)$. Furthermore, it was proved by Kosar, Petrickova, Reiniger and Yeager [41] and independently by Kim, Kwon and Park [36] that for any positive integer k, there is a graph G with $\chi(G^k) < ch(G^k)$. Nevertheless, if G is a cycle, then for any k, the Alon–Tarsi number of G^k equals its chromatic number and hence $\chi(G^k) = ch(G^k)$.

Theorem 4.5.2 *For any cycle C_n and for any positive integer k, $AT(C_n^k) = \chi(C_n^k)$.*

First we calculate the chromatic number of C_n^k. It turns out that it is easier to determine its circular chromatic number.

A (k, d)-*colouring* of a graph G is a mapping $c : V(G) \to \{0, 1, \ldots, k-1\}$ such that for any edge uv of G, $d \leq |c(u) - c(v)| \leq k - d$. The *circular chromatic number* of G is $\chi_c(G) = \min\{k/d : G$ has a (k, d)-colouring$\}$ [68]. As a $(k, 1)$-colouring of G is simply a k-colouring of G, we have $\chi_c(G) \leq \chi(G)$. On the other hand, for any (k, d)-colouring c of G, the mapping $\phi(v) = \lfloor c(v)/d \rfloor$ is a $\lceil \frac{k}{d} \rceil$-colouring of G. So for any graph G, $\chi(G) = \lceil \chi_c(G) \rceil$.

Assume that c is a (k, d)-colouring of G. For $i = 0, 1, \ldots, k-1$, let $X_i = \{v : c(v) \in \{i, i+1, \ldots, i+d-1\}\}$, where the summations are taken modulo k. Then each X_i is an independent set of G and hence $|X_i| \leq \alpha(G)$. On the other hand, each vertex of v is contained in d of the sets X_i. So

$$d|V(G)| = \sum_{i=0}^{k-1} |X_i| \le k\alpha(G).$$

Hence $k/d \ge |V(G)|/\alpha(G)$.

Suppose now $G = C_n^k$ with vertex set $V(G) = \{v_1, v_2, \ldots, v_n\}$, in which $v_i v_j$ is an edge if $|i - j| \le k$ or $|i - j| \ge n - k$. Assume that $n = q(k+1) + p$, where $0 \le p \le k$.

Every set of $k + 1$ consecutive vertices of C_n^k induces a complete graph K_{k+1} and hence contains at most one vertex of an independent set. Therefore G has independence number q and hence $\chi_c(G) \ge n/q$. On the other hand, if $c(v_i) = iq \pmod{n}$, then c is an (n, q)-colouring of G. Indeed, if $v_i \sim v_j$ and $i < j$, then $j - i \le k$ or $j - i \ge n - k$. Hence $q \le |c(v_j) - c(v_i)| \le n - q$. Thus we have $\chi_c(G) = n/q$. So if $n = q(k+1) + p$ where $0 \le p \le k$, then

$$\chi(G) = \lceil n/q \rceil = k + 1 + \lceil p/q \rceil.$$

We shall now prove that $AT(G) = \chi(G)$. The proof is by induction on $\lceil p/q \rceil$.

Case 1 $\lceil p/q \rceil = 0$.

In this case, $\chi(G) = n/q = k + 1$ and we need to prove that $AT(G) = k + 1$. If $q = 1$, then $G = K_{k+1}$ and $k + 1 = \chi(G) \le AT(G) \le col(G) = k + 1$. So assume that $q \ge 2$.

Note that f_G has degree $|E(G)| = kn$. So we need to show that the monomial $\prod_{i=1}^n x_i^k$ is non-vanishing. Let c be the coefficient of $\prod_{i=1}^n x_i^k$ in the expansion of f_G.

It follows from Corollary 4.4.3 that

$$(k!)^n c = \sum_{\sigma} \left((-1)^{\sum_{i=1}^n (k + \sigma(i))} \prod_{i=1}^n \binom{k}{\sigma(i)} \right) f_G(\sigma),$$

where the sum runs over all the mappings $\sigma : \{1, 2, \ldots, n\} \to \{0, 1, \ldots, k\}$. Note that if σ is not a proper colouring of G, then $f_G(\sigma) = 0$. Therefore, we can restrict the sum to those σ which are proper vertex colourings of G with colours $0, 1, \ldots, k$.

The graph G is uniquely $(k + 1)$-colourable. All the $(k + 1)$-colourings of G are obtained from a single $(k + 1)$-colouring of G by permutations of colours. For $i \in \{1, 2, \ldots, n\}$, let $\{i\}_{k+1}$ be the unique integer in $\{0, 1, \ldots, k\}$ congruent to i modulo $(k+1)$. Given a permutation τ of $\{0, 1, \ldots, k\}$, let $\sigma_\tau : V(G) \to \{0, 1, \ldots, k\}$ be defined as follows:

$$\sigma_\tau(v_i) = \tau(\{i\}_{k+1}).$$

A mapping $\sigma : V(G) \to \{0, 1, \ldots, k\}$ is a proper colouring of G if and only if $\sigma = \sigma_\tau$ for some permutation τ of $\{0, 1, \ldots, k\}$. Therefore,

$$(k!)^n c = \sum_{\tau \in S_{k+1}} \left((-1)^{\sum_{i=1}^n (k + \sigma_\tau(v_i))} \prod_{i=1}^n \binom{k}{\sigma_\tau(v_i)} \right) f_G(\sigma_\tau),$$

where S_{k+1} is the set of all permutations of $\{0, 1, \ldots, k\}$.

For $\tau \in S_{k+1}$, for each $j \in \{0, 1, \ldots, k\}$, there are exactly q indices $i \in \{1, 2, \ldots, n\}$ with $\sigma_\tau(i) = j$. So

$$(-1)^{\sum_{i=1}^n (k + \sigma_\tau(i))} = (-1)^{\frac{qk(k+1)}{2}}, \text{ and}$$

$$\prod_{i=1}^n \binom{k}{\sigma_\tau(i)} = \prod_{i=0}^k \binom{k}{i}^q.$$

Therefore $(-1)^{\sum_{i=1}^n (k + \sigma_\tau(v_i))} \prod_{i=1}^n \binom{k}{\sigma_\tau(v_i)}$ is a non-zero constant and $c \neq 0$ if and only if $\sum_{\tau \in S_{k+1}} f_G(\sigma_\tau) \neq 0$.

We shall show that $f_G(\sigma_\tau)$ is also a constant independent of τ. We view the cycle C_n as embedded in the plane as a circle, so that the vertices $v_1, v_2, \ldots, v_n$ of G are ordered cyclically. Each vertex v of G has k neighbours to its left and k neighbours to its right. Assume that D is the orientation of G in which each edge $v_i v_j$ is oriented from v_i to v_j if v_i is to the left of v_j in the cyclic order. Assume that $f_G(\mathbf{x}) = \prod_{(u,v) \in A(D)} (x_u - x_v)$. We consider each pair of colours $i > j$. Each vertex of colour i is adjacent to two vertices of colour j, one to its left, and one to its right. These two edges incident to this vertex of colour i contributes $(i - j)(j - i) = -(i - j)^2$ to $f_G(\sigma_\tau)$. There are q vertices of colour i. So the contribution to $f_G(\sigma_\tau)$ from the pair of colours $i > j$ is $(-(i - j)^2)^q$. This contribution is independent of τ. So $f_G(\sigma_\tau)$ is a constant independent of τ. Hence $\prod_{i=1}^n x_i^k$ is a non-vanishing monomial of f_G and $AT(G) \leq k + 1$. This completes the proof of Case 1.

Case 2 $\lceil p/q \rceil = 1$.

In this case, G is a subgraph of $C^{k+1}_{q(k+2)}$. Indeed, let the mapping $\phi : \{1, 2, \ldots, q(k+1)+p\} \to \{1, 2, \ldots, q(k+2)\}$ be defined as

$$\phi(i) = \begin{cases} i + \lfloor \frac{i}{k} \rfloor, & \text{if } i < (q-p)k \\ i + (q-p), & \text{otherwise.} \end{cases}$$

Then ϕ is an embedding of $G = C^k_{q(k+1)+p}$ into $C^{k+1}_{q(k+2)}$. By Case 1, $AT(C^{k+1}_{q(k+2)}) \leq k+2$. By Lemma 3.1.7, $AT(G) \leq k+2$.

Case 3 $\lceil p/q \rceil \geq 2$.

In this case, $n = q(k+1) + p = q(k+2) + (p-q)$, where $1 \leq p - q \leq k - q$. Hence $\chi(C^{k+1}_n) = (k+2) + \lceil (p-q)/q \rceil = \chi(G)$. By induction hypothesis $AT(C^{k+1}_n) = \chi(C^{k+1}_n)$. Therefore $AT(G) = \chi(G)$. ∎

4.6 Alon–Tarsi numbers of toroidal grids

Let $n, m \geq 3$ be integers. The Cartesian product $C_n \square C_m$ of two cycles is called a *toroidal grid*, and we denote it by $T_{n,m}$. Note that $T_{n,m}$ is a 4-regular graph. If both n, m are even, then $T_{n,m}$ is a bipartite graph. It follows from Corollary 3.2.2 that $AT(T_{n,m}) = 3$. On the other hand, by Theorem 3.5.3, for any integers n, m, $AT(T_{n,m}) \leq 4$. Of course, $AT(T_{n,m}) \geq 3$ for any integers $n, m \geq 3$ (any orientation of $T_{n,m}$ has maximum out-degree at least 2). The problem is to determine whether $AT(T_{n,m})$ is 3 or 4, when at least one of n, m is odd. This problem is solved in [45].

Theorem 4.6.1 *If both n, m are odd, then $AT(T_{n,m}) = 4$. If one of n, m is odd and the other is even, then $AT(T_{n,m}) = 3$.*

Observe that $P_n \square C_m$ is a subgraph of $T_{n',m}$ for any $n' \geq n$. Hence $AT(P_n \square C_m) \leq AT(T_{2n,m}) = 3$ for $n \geq 2$, which was proved in Section 3.3.

The proof of Theorem 4.6.1 uses the concept of anti-Hermitian matrix and basic properties of complex numbers and matrices.

Definition 4.6.2 *The* conjugation $\bar{z}$ *of a complex number* $z = a + bi$ *is defined as* $\bar{z} = a - bi$, *where* a, b *are real numbers. The* conjugate transpose $\overline{M}$ *of a complex matrix* M *is defined as* $\overline{M}_{i,j} = \overline{M_{j,i}}$. *(Here* $\overline{M}_{i,j}$ *is the* (i,j)*th entry of the matrix* $\overline{M}$, *and* $\overline{M_{j,i}}$ *is the conjugate of the complex number* $M_{j,i}$.) *A square complex matrix* M *is* Hermitian *if* $\overline{M} = M$ *and* M *is* anti-Hermitian *(or* skew Hermitian*) if* $\overline{M} = -M$.

The following lemma lists some basic properties of conjugate transposes. These properties can be easily verified from the definitions.

Lemma 4.6.3 *Let* A, B *be complex matrices,* **x** *a complex vector and* z, z' *are complex numbers. Then the following hold:*

1. $\overline{\overline{A}} = A, \overline{\overline{z}} = z.$

2. $\overline{(A + B)} = \overline{A} + \overline{B}, \overline{(z + z')} = \overline{z} + \overline{z'}.$

3. $\overline{(zA)} = \overline{z}\overline{A}, \overline{(1/z)} = 1/\overline{z}.$

4. $\overline{(AB)} = \overline{B}\ \overline{A}, \overline{(zz')} = \overline{z}\ \overline{z'}.$

Note that when we write the sum of two matrices, they are assumed to be of the same sizes; when we write the product AB of two matrices, the number of columns in A is assumed to be equal to the number of rows in B. We view an n-dimensional complex vector **x** as an $n \times 1$ matrix. So $\overline{\mathbf{x}}$ is a $1 \times n$ matrix.

Definition 4.6.4 *Assume that* **u**, **v** *are* n-dimensional complex *vectors. Then* $\overline{\mathbf{u}}\mathbf{v}$ *is called the* inner *product of* **u** *and* **v**. *The* norm *of a complex number* $z = a + bi$ *is* $|z| = \sqrt{a^2 + b^2} = \sqrt{z\overline{z}}$ *and the norm of a complex vector* **u** *is* $|\mathbf{u}| = \sqrt{\overline{\mathbf{u}}\mathbf{u}}$.

Note that the inner product of two complex vectors is a complex number.

Lemma 4.6.5 *Assume that* d *is a positive integer and that* H *is a* $2d$-regular graph with vertex set $\{v_1, v_2, \ldots, v_n\}$. *Let* $G = H \square C_k$ *and set* $N = nk$. *Fix a field* $\mathbb{F}$ *and a subset* $A \subseteq \mathbb{F}$ *of size* $|A| = d+2$. *Let* $\mathcal{U}$ *be the set of all proper colourings of* H *using colours from* A. *Each element of* $\mathcal{U}$ *is a vector* $\mathbf{u} = (u_1, u_2, \ldots, u_n)$, *where* $u_i \in A$ *is the colour assigned to vertex* v_i. *For* $\mathbf{u}, \mathbf{u}' \in \mathcal{U}$, *let*

$$M_{\mathbf{u},\mathbf{u}'} = f_H(u_1, \ldots, u_n) \cdot \prod_{i=1}^{n} \frac{u_i - u_i'}{\prod_{b \in A - \{u_i\}}(u_i - b)}.$$

Let $M = [M_{u,u'}]$ be the matrix whose rows and columns are indexed by elements of $\mathcal{U}$. Then

$$c_{G,d+1} = tr M^k.$$

Proof. By the coefficient formula in Theorem 4.3.1,

$$c_{G,d+1} = \sum_{(a_1,\ldots,a_N) \in A^N} \frac{f_G(a_1,\ldots,a_N)}{\prod_{i=1}^{N} \prod_{b \in A - \{a_i\}}(a_i - b)}, \quad (4.6)$$

where $c_{G,d+1} = c_{G,\|S\|}$ with $|S_1| = \ldots = |S_n| = d + 2$. We may restrict the summation to those $(a_1,\ldots,a_N) \in A^N$ that are proper colourings of G (for otherwise $f_G(a_1,\ldots,a_N) = 0$). Each proper colouring $(a_1,\ldots,a_N)$ of G corresponds to a sequence $\mathbf{u}^1, \mathbf{u}^2, \ldots, \mathbf{u}^k$ of proper colourings of H, where $\mathbf{u}^i$ is the colouring of the ith copy of H, i.e., colouring of those vertices of $G \times C_k$ whose second coordinate is v_i (the ith vertex of C_k). Let $\mathbf{u}^i = (u_{i,1}, u_{i,2}, \ldots, u_{i,n})$. We denote by $f_H(\mathbf{u}_i) = f_H(u_{i,1}, u_{i,2}, \ldots, u_{i,n})$. Then

$$f_G(a_1,\ldots,a_N) = \prod_{i=1}^{k} f_H(\mathbf{u}^i) \cdot \prod_{i=1}^{k} \prod_{j=1}^{n}(u_{i+1,j} - u_{i,j}).$$

In the summation above, $u_{k+1,j} = u_{1,j}$. Therefore

$$\frac{f_G(a_1,\ldots,a_N)}{\prod_{i=1}^{N} \prod_{b \in A - \{a_i\}}(a_i - b)} = M_{\mathbf{u}^1,\mathbf{u}^2} \cdot M_{\mathbf{u}^2,\mathbf{u}^3} \cdot \ldots \cdot M_{\mathbf{u}^k,\mathbf{u}^1}.$$

The summation of all these products is exactly the trace of the matrix M^k. ∎

The matrix M depends on A and H. Assume that $H = C_{2s+1}$. Let $A = \{1, w, w^2\}$, where $w = e^{2\pi i/3}$ is a cube root of 1. We shall show that in this case, the trace of M^k is non-zero if k is even, and trace of M^k is zero if k is odd.

Lemma 4.6.6 *The matrix M is an anti-Hermitian matrix.*

Proof. In this proof, the indices are modulo $2s + 1$. Let $\mathbf{u} = (u_1, u_2, \ldots, u_{2s+1})$ be a proper 3-colouring of C_{2s+1} with colour set $A = \{1, w, w^2\}$. Let u_i^* be the unique element in $A - \{u_i, u_{i-1}\}$. Then

$$\frac{u_i - u_{i-1}}{\prod_{b \in A - \{u_i\}}(u_i - b)} = \frac{1}{u_i - u_i^*}.$$

As $f_H(u_1, \ldots, u_{2s+1}) = \prod_{i=1}^{2s+1}(u_i - u_{i-1})$, we conclude that

$$
\begin{aligned}
M_{\mathbf{u},\mathbf{u}'} &= f_H(u_1, \ldots, u_{2s+1}) \cdot \prod_{i=1}^{2s+1} \frac{u_i - u_i'}{\prod_{b \in A - \{u_i\}}(u_i - b)} \\
&= \prod_{i=1}^{2s+1} \frac{(u_i - u_i')(u_i - u_{i-1})}{\prod_{b \in A - \{u_i\}}(u_i - b)} \\
&= \prod_{i=1}^{2s+1} \frac{u_i - u_i'}{u_i - u_i^*}.
\end{aligned}
$$

We shall now show that

$$
M_{\mathbf{u},\mathbf{u}'} = -\overline{M_{\mathbf{u}',\mathbf{u}}}. \tag{4.7}
$$

If $u_i = u_i'$ for some i, then $M_{\mathbf{u},\mathbf{u}'} = 0$. So (4.7) holds. Assume therefore that $u_i \neq u_i'$ for all i. Then (4.7) is equivalent to the following:

$$
\prod_{i=1}^{2s+1} \frac{(u_i - u_i')(\overline{u}_i - \overline{u}_i^*)}{|u_i - u_i^*|^2} = -\prod_{i=1}^{2s+1} \frac{(\overline{u}_i' - \overline{u}_i)(u_i' - u_i'^*)}{|u_i' - u_i'^*|^2}. \tag{4.8}
$$

Since $|u_i' - u_i'^*|^2 = |u_i - u_i^*|^2 = 3$, and $\overline{z} = \frac{1}{z}$ for $z \in \{u_i, u_i', u_i^*\}$, (4.8) is equivalent to

$$
\prod_{i=1}^{2s+1} (u_i - u_i') \frac{u_i^* - u_i}{u_i u_i^*} = -\prod_{i=1}^{2s+1} (u_i' - u_i'^*) \frac{u_i - u_i'}{u_i u_i'}. \tag{4.9}
$$

Now (4.9) is equivalent to

$$
\prod_{i=1}^{2s+1} \frac{u_i^* - u_i}{u_i^*} = -\prod_{i=1}^{2s+1} \frac{u_i' - u_i'^*}{u_i'}. \tag{4.10}
$$

Let $\epsilon_i = \frac{u_i}{u_{i-1}}$ and $\delta_i = \frac{u_i'}{u_{i-1}'}$. Then

$$
\epsilon_i, \delta_i \in \{w, w^2\}, \quad u_i^* = u_i \epsilon_i, \quad u_i'^* = u_i' \delta_i.
$$

Hence

$$
\frac{u_i}{u_i^*} = \overline{\epsilon}_i, \quad \frac{u_i'^*}{u_i'} = \delta_i.
$$

We re-write (4.10) as

$$
\prod_{i=1}^{2s+1} (1 - \overline{\epsilon}_i) = -\prod_{i=1}^{2s+1} (1 - \delta_i).
$$

As

$$\prod_{i=1}^{2s+1} \epsilon_i = \prod_{i=1}^{2s+1} \delta_i = 1,$$

(4.10) is equivalent to

$$\prod_{i=1}^{2s+1} (1 - \delta_i) = -\prod_{i=1}^{2s+1} \epsilon_i(1 - \overline{\epsilon}_i) = \prod_{i=1}^{2s+1} (1 - \epsilon_i). \quad (4.11)$$

Note that $w = -\frac{1}{2} + \frac{\sqrt{3}}{2}i$ and $w^2 = -\frac{1}{2} - \frac{\sqrt{3}}{2}i$. So $1 - w = \sqrt{3}iw^2$ and $1 - w^2 = -\sqrt{3}iw$. So for $\epsilon \in \{w, w^2\}$, $1 - \epsilon = \pm i\sqrt{3}\epsilon^2$ and the signs are different for w and w^2. As $\prod_{i=1}^{2s+1} \epsilon_i^2 = \prod_{i=1}^{2s+1} \delta_i^2 = 1$, we conclude that (4.11) holds if and only if

$$|\{i : \epsilon_i = w\}| \equiv |\{i : \delta_i = w\}| \pmod 2. \quad (4.12)$$

Colour i with white if $u_i = u_i'w$, and colour i with black if $u_i = u_i'w^2$. Then $\epsilon_i = \delta_i$ if i and $i - 1$ have the same colour, and $\epsilon_i \neq \delta_i$ if i and $i - 1$ have distinct colours. As there is an even number of indices i for which i and $i - 1$ are of distinct colours, so there is an even number of indices i for which $\epsilon_i \neq \delta_i$. Therefore (4.12) holds and hence M is an anti-Hermitian matrix. ∎

Recall that a complex number λ is *pure imaginary* if its real part is zero, which is equivalent to $\lambda = -\overline{\lambda}$.

Lemma 4.6.7 *If M is an anti-Hermitian matrix, then all its eigenvalues are pure imaginary.*

Proof. Assume λ is an eigenvalue of M and $\mathbf{x}$ is the corresponding eigenvector. Then $\overline{\mathbf{x}}M\mathbf{x} = \lambda|\mathbf{x}|^2$. Hence

$$\overline{\lambda}|\mathbf{x}|^2 = \overline{\overline{\mathbf{x}}M\mathbf{x}} = \mathbf{x}\overline{M}\overline{\mathbf{x}} = -\overline{\mathbf{x}}M\mathbf{x} = -\lambda|\mathbf{x}|^2.$$

Thus $\lambda = -\overline{\lambda}$ and hence λ is pure imaginary. ∎

Note that the same proof shows that if M is a Hermitian matrix, then all its eigenvalues are real.

Now we are ready to complete the proof of Theorem 4.6.1. Assume that $G = C_{2s+1} \square C_k$. Let M be the matrix defined as above. Note that the matrix M is a non-zero matrix. If k is even, then since the eigenvalues of M are all imaginary, the eigenvalues of M^k are all real numbers, and if $k \equiv 0 \pmod 4$, then all the eigenvalues are non-negative; if $k \equiv 2 \pmod 4$, then all the eigenvalues are non-positive. Since M is non-zero, M^k has at least one non-zero eigenvalue. Hence $tr M^k \neq 0$ (recall that the trace of a square matrix is the summation of all its eigenvalues). By Lemma 4.6.7, $c_{G,2} \neq 0$ and hence $AT(G) = 3$. If k is odd, then the eigenvalues of M^k are also all pure imaginary. Thus $tr M^k$ is a pure imaginary number. On the other hand, $tr M^k = c_{G,2}$ is a real number. Hence $tr M^k = c_{G,2} = 0$. So $AT(G) \geq 4$. As observed above, $AT(G) \leq 4$. So equality $AT(G) = 4$ holds.   ∎

Corollary 4.6.8 *Assume that* $n, m \geq 2$ *are integers. If* nm *is even, then* $ch(T_{n,m}) = 3$. *If* nm *is odd, then* $3 \leq ch(T_{n,m}) \leq 4$.

It was conjectured in [14] that $ch(T_{n,m}) = 3$ for all $n, m \geq 3$. It remains open whether the conjecture holds when n, m are odd.

4.7   List colouring of line graphs

Although there are graphs whose choice numbers are much larger than their chromatic numbers (cf: Theorem 1.0.15), there are many graphs whose choice numbers equal their chromatic numbers. Such graphs are called *chromatic choosable*. The results in the previous section show that $K_{2 \star n}$ and cycle powers C_n^k are chromatic choosable. The problem as to which graphs are chromatic choosable has been studied a lot in the literature. One of the most well-known problems is the so-called list colouring conjecture, which was suggested independently by many researchers, including Vizing, Gupta, Allbertson, Collins and Tucker.

Conjecture 4.7.1 (List colouring Conjecture) *For every graph* G *(parallel edges are allowed), the list edge-chromatic number* $\chi_l'(G)$ *equals its edge chromatic number* $\chi'(G)$.

Conjecture 4.7.1 is equivalent to the following conjecture, which is the restriction of Conjetcure 4.7.1 to r-edge colourable r-regular graphs.

Conjecture 4.7.2 (List colouring Conjecture) *Every* r-edge *colourable* r-regular graph G *is* r-edge choosable.

Lemma 4.7.3 *Conjecture 4.7.1 and Conjecture 4.7.2 are equivalent.*

Proof. It is obvious that Conjecture 4.7.1 implies Conjecture 4.7.2. To see the converse implication, we assume that Conjecture 4.7.2 holds and G is an arbitrary r-edge colourable graph. Let H be obtained from two vertex disjoint copies of G, say G and G', where the vertex set of G' is $V' = \{v' : v \in V(G)\}$, by adding $r - d_G(v)$ parallel edges between v and v' for each vertex v of G. Then H is r-regular and r-edge colourable. Since Conjecture 4.7.2 is assumed to be true for all r-edge colourable r-regular graphs, H is r-edge choosable. Hence G is also r-edge choosable. As G is an arbitrary r-edge colourable graph, Conjecture 4.7.1 is true. ∎

Let G be an n-vertex r-regular r-edge colourable graph. Let $L(G)$ be the line graph of G. We are interested in the Alon–Tarsi number of $L(G)$. In particular, we are interested in the problem whether $L(G)$ has Alon–Tarsi number r. As a strengthening of Conjecture 4.7.2, one may wonder if it is true that for any r-edge colourable r-regular graph G, its line graph $L(G)$ has Alon–Tarsi number r. However, this is not true. For example, if $G = K_{3,3}$, then G is 3-edge colourable and 3-regular, but its line graph $L(G)$ has Alon–Tarsi number 4. Nevertheless, for some classes of r-edge colourable r-regular graphs G, $L(G)$ does have Alon–Tarsi number r. In the next two sections, we shall show that the following hold:

(1) If G is an r-edge colourable r-regular planar graph, then $L(G)$ has Alon–Tarsi number r.

(2) If $G = K_{p+1}$ and p is an odd prime, then $L(G)$ has Alon–Tarsi number p.

Let D be an orientation of $L(G)$ and let

$$f_D(\mathbf{x}) = \prod_{(e,e') \in E(D)} (x_e - x_{e'})$$

be the graph polynomial of $L(G)$ associated with D. Note that for a different orientation D' of $L(G)$, $f_D(\mathbf{x}) = \pm f_{D'}(\mathbf{x})$. We are only interested in whether a certain monomial in f_D is non-vanishing or not. So the choice of the orientation D is irrelevant. However, for some of the proofs, we may choose appropriate orientations D to make the calculations easier.

The line graph $L(G)$ is $2(r-1)$-regular (if e, e' are parallel edges in G, then e and e' are connected by two parallel edges in $L(G)$). Thus the number of variables x_e is $|E(G)| = nr/2$, and the degree of f_D is $|E(L(G))| = nr(r-1)/2$. Thus the Alon–Tarsi number of $L(G)$ is r if and only if the monomial $\prod_{e \in E(G)} x_e^{r-1}$ is non-vanishing.

By Corollary 4.4.3, the coefficient of the monomial $\prod_{e \in E(G)} x_e^{r-1}$ in $f_D(\mathbf{x})$ is equal to

$$((r-1)!)^{-nr/2} \sum_{\phi : E(G) \to \{0,1,\ldots,r-1\}}$$

$$\times \left(\prod_{e \in E(G)} (-1)^{(r-1+\phi(e))} \binom{r-1}{\phi(e)} \right)$$

$$\times f_D(\phi).$$

If $\phi : E(G) \to \{0,1,\ldots,r-1\}$ is not a proper edge colouring of G, then $f_D(\phi) = 0$. Thus the summation above can be restricted to proper edge colourings ϕ of G. On the other hand, if $\phi : E(G) \to \{0,1,\ldots,r-1\}$ is a proper edge colouring of G, then $\prod_{e \in E(G)}(-1)^{(r-1+\phi(e))} \binom{r-1}{\phi(e)}$ is a constant, as there are exactly $n/2$ edges e for which $\phi(e) = i$ for each $i \in \{0,1,\ldots,r-1\}$. Therefore, to prove that the coefficient of the monomial $\prod_{e \in E(G)} x_e^{r-1}$ in $f_D(\mathbf{x})$ is non-zero, it is equivalent to showing that $\sum_{\phi \in \mathcal{C}} f_D(\phi) \neq 0$, where $\mathcal{C}$ is the set of mappings $\phi : E(G) \to \{0,1,\ldots,r-1\}$ that are proper edge colourings of G.

For a vertex v of G, let $E(v)$ be the set of edges of G incident to v, and denote by Q_v the r-clique of $L(G)$ induced by $E(v)$. Let

$$f_{D,v}(\mathbf{x}) = \prod_{(e,e') \in E(D), e, e' \in E(Q_v)} (x_e - x_{e'}).$$

Then $f_D(\mathbf{x}) = \prod_{v \in V(G)} f_{D,v}(\mathbf{x})$. If ϕ is a proper edge colouring of G, then for any vertex v of G,

$$|f_{D,v}(\phi)| = \prod_{0 \leq i < j \leq r-1} (j - i), \text{ and hence } |f_D(\phi)|$$

$$= \left(\prod_{0 \leq i < j \leq r-1} (j - i) \right)^n.$$

For $\phi \in \mathcal{C}$, let $sign_D(\phi) = \pm 1$ (respectively, $sign_{D,v}(\phi) = \pm 1$), depending on whether $f_D(\phi)$ (respectively, $f_{D,v}(\phi)$) is positive or negative. As $|f_D(\phi)|$ is a constant, to show that the coefficient of $\prod_{e \in E(G)} x_e^{r-1}$ in $f_D(\mathbf{x})$ is non-zero, it is equivalent to showing that $\sum_{\phi \in \mathcal{C}} sign_D(\phi) \neq 0$. So we have the following lemma.

Lemma 4.7.4 *Assume that G is an r-edge colourable r-regular graph and D is an orientation of $L(G)$. Then $L(G)$ has Alon–Tarsi number r if and only if $\sum_{\phi \in C} sign_D(\phi) \neq 0$, where C is the set of proper r-edge colourings of G.*

4.8 r-regular planar graphs

In this section, we prove that if G is an r-edge colourable r-regular planar graph, then $L(G)$ has Alon–Tarsi number r. This result was proved by Ellingham and Goddyn [22]. The proof presented below is an unpublished proof given by the first author.

By Lemma 4.7.4, it suffices to show that for a fixed orientation D of $L(G)$, $\sum_{\phi \in C} sign_D(\phi) \neq 0$.

For this purpose, we prove a stronger result:

Lemma 4.8.1 *There is an orientation D of $L(G)$ such that for all $\phi \in C$, $sign_D(\phi) = (-1)^{\binom{r-1}{2}n/2}$, a non-zero constant.*

Proof. Assume that D, D' are distinct orientations of $L(G)$. If D and D' differ in an even number of edges, then $f_{D'}(\mathbf{x}) = f_D(\mathbf{x})$. Otherwise $f_{D'}(\mathbf{x}) = -f_D(\mathbf{x})$.

Let M be a one-factor of G. We define an orientation of $L(G)$ as follows: For each vertex v of G, order the edges in $E(v)$ as $(e_1, e_2, \ldots, e_r)$, where $e_1 \in M$ and the edges occur clockwise in this order in a plane embedding of G. Let Q_v be the $r-$ clique of $L(G)$ induced by the r edges at the vertex v of G. We orient the edges in Q_v as $e_i \to e_j$ where $1 \leq i < j \leq r$. The union of the orientations of all the cliques $\{Q_v : v \in V(G)\}$ gives an orientation of $L(G)$, which we denote by D_M. Note that for $e = uv$, the index of e in $E(v)$ and in $E(u)$ may be distinct.

Claim 4.8.2 *If M and M' are both one-factors of G, then D_M and $D_{M'}$ differ in an even number of edges. Consequently $f_{D_M}(\mathbf{x}) = f_{D_{M'}}(\mathbf{x})$ and $sign_{D_M}(\phi) = sign_{D_{M'}}(\phi)$ for $\phi \in C$.*

Proof. For each vertex v of G, let α_v be the number of edges in Q_v for which D_M and $D_{M'}$ orient differently. We need to show that $\sum_{v \in V(G)} \alpha_v \equiv 0 \pmod{2}$.

Let $e_{M,v}$ and $e_{M',v}$ be the edges of M and M' incident to v, respectively. If $e_{M,v} = e_{M',v}$, then $\alpha_v = 0$. Otherwise, let

$[e_{M,v}, e_{M',v})$ be the set of edges in $E(v)$ from $e_{M,v}$ to $e_{M',v}$ (include $e_{M,v}$ but exclude $e_{M',v}$) in clockwise direction, and define the set $[e_{M',v}, e_{M,v})$ of edges similarly. It is easy to see that D_M and $D_{M'}$ orient the edge ee' in Q_v differently if and only if one of e, e' lies in $[e_{M,v}, e_{M',v})$, and the other lies in $[e_{M',v}, e_{M,v})$. Thus $\alpha_v = |[e_{M,v}, e_{M',v})| \times |[e_{M',v}, e_{M,v})|$. If r is odd, then α_v is even for every v and hence $\sum_{v \in V(G)} \alpha_v \equiv 0 \pmod{2}$.

Assume that r is even. Then for each vertex v, $\alpha_v \equiv |[e_{M,v}, e_{M',v})| \equiv |[e_{M',v}, e_{M,v})| \pmod{2}$. Let $C = (v_1, v_2, \ldots, v_{2k})$ be an even cycle of $M \cup M'$. The number of edges in $E(v_i)$ contained in the interior of C is $|[e_{M,v}, e_{M',v})| - 1$ or $|[e_{M',v}, e_{M,v})| - 1$, and hence has the same parity as $\alpha_v - 1$. Since each vertex in the interior of C has even degree r, by handshaking lemma, the total number of edges in the interior with one end on cycle C is even. Thus $\sum_{i=1}^{2k} \alpha_{v_i} \equiv \sum_{i=1}^{2k} (\alpha_{v_i} - 1) \equiv 0 \pmod{2}$. As $(M \cup M') - (M \cap M')$ is the disjoint union of even cycles, we conclude that $\sum_{i=1}^{2k} \alpha_{v_i} \equiv 0 \pmod{2}$. $\blacksquare$

We shall prove that if M is a one-factor of G, then for any $\phi \in \mathcal{C}$, $sign_{D_M}(\phi) = (-1)^{\binom{r-1}{2} n/2}$. By Claim 4.8.2, we may assume that $M = \phi^{-1}(r-1)$ consists of edges of colour $r-1$.

Let G' be obtained from G by contracting all edges of colour $r-1$ (and all the other edges coloured as in G). For each colour $i \in \{0, 1, \ldots, r-2\}$, the subgraph of G' induced by edges of colour i induce a 2-factor. We say *monochromatic cycles of colours i and j cross at u* if the cyclic order of the colours of the i- j-coloured edges incident to u is i, j, i, j. Denote by β_u the number of pairs of monochromatic cycles of G' that cross at u.

For $u \in V(G')$, let $e_u = x_u y_u$ be the edge of G whose contraction results in u. For each pair of colours $0 \le i, j \le r-2$, in each of $P_{D_M, x_u}(\phi)$ and $P_{D_M, y_u}(\phi)$, $(j - i)$ or $(i - j)$ occurs as a factor. If cycles of colours i and j cross at u, then either $(j - i)$ or $(i - j)$ occurs twice in the product $P_{D_M, x_u}(\phi) P_{D_M, y_u}(\phi)$. If cycles of colours i and j do not cross at u, then each of $(j - i)$ and $(i - j)$ occurs once in $P_{D_M, x_u}(\phi) P_{D_M, y_u}(\phi)$. Note that $\binom{r-1}{2} - \beta_u$ is the number of pairs i, j for which monochromatic cycles of colours i and j do not cross. Therefore,

$$sign_{D_M, x_u}(\phi) sign_{D_M, y_u}(\phi) = (-1)^{\binom{r-1}{2} - \beta_u}.$$

Since every pair of monochromatic cycles cross each other an even number of times, we have $\sum_{u \in V(G')} \beta_u = 0 \pmod{2}$.

Therefore,

$$sign_{D_M}(\phi) = \prod_{u \in V(G')} sign_{D_M,x_u}(\phi)sign_{D_M,y_u}(\phi) = (-1)^{\binom{r-1}{2}n/2}.$$

This completes the proof of Lemma 4.8.1, as well as Theorem 4.8.3. ∎

Corollary 4.8.3 *If G is an r-regular r-edge colourable planar graph, then $L(G)$ has Alon–Tarsi number r, and hence G is r-edge paintable as well as r-edge choosable.*

4.9   Complete graphs K_{p+1} for odd prime p

In this section, we assume that p is an odd prime and $G = K_{p+1}$. It is well-known [7, 63] and easy to verify that G is p-edge colourable. We shall prove that the line graph $L(G)$ has Alon–Tarsi number p, and hence G is p-edge paintable as well as p-edge choosable. This result was proved by Schauz [56], and the proof presented below is a shorter proof adapted from that proof. The proof is still quite long and complicated. Two polynomials are introduced so that one can apply Corollary 4.4.3 to these polynomials to derive the desired inequality.

Theorem 4.9.1 *If $G = K_{p+1}$, p is an odd prime, then $AT(L(G)) = p$.*

Since $\chi(L(G)) = p$, we know that $AT(L(G)) \geq p$. It remains to show that $AT(L(G)) \leq p$.

Let D be an orientation of $L(G)$. We shall show that the coefficient of the monomial $\prod_{e \in E(G)} x_e^{p-1}$ in the expansion of the graph polynomial $f_D(\mathbf{x})$ is non-zero.

Let $\mathcal{C}$ be the set of proper edge colourings of G, using colours in $\{0, 1, \ldots, p-1\}$. By Lemma 4.7.4, it suffices to show that

$$\sum_{\phi \in \mathcal{C}} sign_D(\phi) \neq 0. \tag{4.13}$$

To prove inequality 4.13, we shall reduce this inequality to simpler inequalities in a few steps. The first step is to show that we may reduce the summation from $\phi \in \mathcal{C}$ to $\phi \in \mathcal{C}'$, where $\mathcal{C}'$ consists of those proper edge colourings ϕ of $L(G)$ for which $\phi(pi) = i$ for $i \in \{0, 1, \ldots, p-1\}$.

We observed before, in the definiton of $f_D(\mathbf{x})$, the orientation D of $L(G)$ is not important: for two different orientations D and D' of $L(G)$, $f_D(\mathbf{x}) = \pm f_{D'}(\mathbf{x})$. However, for our convenience, we shall choose D carefully.

We may assume that the vertices of G are $\{0, 1, \ldots, p\}$. For each vertex $i \in \{0, 1, \ldots, p\}$, let $E_i = \{ij : j \in \{0, 1, \ldots, p\}, j \neq i\}$ be the set of edges incident to i.

We order the edges in E_p as $p0 < p1 < \ldots < p(p-1)$ and for $i \in \{0, 1, \ldots, p-1\}$, we order the edges in E_i as $i0 < i1 < \ldots < i(i-1) < ip < i(i+1) < \ldots < i(p-1)$. Let D be the orientation of $L(G)$ in which an edge $ee' \in E(L(G))$ is oriented as (e, e') if $e < e'$ in this ordering.

Assume that $\phi \in \mathcal{C}$. Then let

$$\phi_p = (\phi(p0), \phi(p1), \ldots, \phi(p(p-1)))$$

and for $i = 0, 1, \ldots, p-1$, let

$$\phi_i = (\phi(i0), \phi(i1), \ldots, \phi(i(i-1)), \phi(ip), \phi(i(i+1)), \ldots, \phi(i(p-1))).$$

Each ϕ_i is a permutation of $\{0, 1, \ldots, p-1\}$.

For a permutation π of $\{0, 1, \ldots, p-1\}$, an *inversion* of π is a pair of indices (i, j) such that $i < j$ and $\pi(i) > \pi(j)$. We say π is an *even* (respectively, *odd*) permutation if π has an even number (respectively, an odd number) of inversions. The *sign* sign(π) of π is 1 (respectively, -1) if π is an even permutation (respectively, an odd permutation). If $1 \leq i < j \leq p-1$, $\pi'(k) = \pi(k)$ for $k \neq i, j$, $\pi'(i) = \pi(j)$ and $\pi'(j) = \pi(i)$, then we say π' is obtained from π by interchanging two entries. Note that interchanging two entries of a permutation changes the parity of the permutation.

It follows from the definition that

$$\text{sign}_D(\phi) = \prod_{i=0}^{p} \text{sign}(\phi_i).$$

Lemma 4.9.2 *Assume that $\phi, \psi \in \mathcal{C}$. If there is a permutation π of $GF(p)$ such that for every edge e of G, $\psi(e) = \pi(\phi(e))$, then* $\text{sign}_D(\phi) = \text{sign}_D(\psi)$.

Proof. It suffices to consider the case when $\pi = (ab)$ is an involution that interchanges two colours a, b. In this case, for each vertex i, ψ_i is obtained from ϕ_i by interchanging two entries. Hence $\text{sign}(\phi_i) = -\text{sign}(\psi_i)$. As there is an even number of vertices, we have $\text{sign}_D(\phi) = \text{sign}_D(\psi)$. ∎

Let $\mathcal{C}'$ be the set of proper edge colourings ϕ of $L(G)$ for which $\phi(pi) = i$ for $i \in \{0, 1, \ldots, p-1\}$. Then any colouring $\psi \in \mathcal{C}$ is obtained from a colouring $\phi \in \mathcal{C}'$ by a permutation of colours, and each permutation $\phi \in \mathcal{C}'$ corresponds to $p!$ colourings in $\mathcal{C}$. Then by Lemma 4.9.2,

$$\sum_{\phi \in \mathcal{C}} sign_D(\phi) = p! \sum_{\phi \in \mathcal{C}'} sign_D(\phi).$$

Thus to prove that $L(G)$ has Alon–Tarsi number p, it suffices to show that

$$\sum_{\phi \in \mathcal{C}'} sign_D(\phi) \neq 0. \qquad (4.14)$$

By definition, for all $\phi \in \mathcal{C}'$, $\text{sign}(\phi_p) = 1$. So

$$\text{sign}_D(\phi) = \prod_{i=0}^{p-1} \text{sign}(\phi_i).$$

For convenience, we do not distinguish between a permutation of $\{0, 1, \ldots, p-1\}$ and a vector whose coordinates form the permutation.

For $\phi \in \mathcal{C}'$, let A_ϕ be the $p \times p$ matrix whose row vectors are $\phi_0, \phi_1, \ldots, \phi_{p-1}$, i.e.,

$$A_\phi = \begin{pmatrix} \phi(0p) & \phi(01) & \phi(02) & \cdots & \phi(0(p-1)) \\ \phi(10) & \phi(1p) & \phi(12) & \cdots & \phi(1(p-1)) \\ \vdots & \vdots & \vdots & \cdots & \vdots \\ \phi((p-1)0) & \phi((p-1)1) & \phi((p-1)2) & \cdots & \phi((p-1)p) \end{pmatrix}.$$

Note that the diagonal vector of A_ϕ is

$$(\phi(0p), \phi(1p), \ldots, \phi((p-1)p)).$$

In the following, we assume that p is an odd prime, and all the additions are carried out in $GF(p)$. For convenience, for a $p \times p$ matrix A over $GF(p)$, the rows and columns of A are indexed as the 0-th row, the 1st row, ..., the $(p-1)$th row, 0-th column, the 1st column, ..., the $(p-1)$th column. So the entries of A are $\{A_{i,j} : 0 \leq i \leq p-1, 0 \leq j \leq p-1\}$.

Definition 4.9.3 *We denote by $\mathcal{L}_p$ the set of $p \times p$ matrices over $GF(p)$ for which the following hold:*

1. $A_{i,j} \neq A_{i,j'}$ *if $j \neq j'$.*

2. $A_{i,j} = A_{j,i}$ *for any i, j.*

3. $A_{i,i} = i$.

Note that each $A \in \mathcal{L}_p$ is a symmetric Latin square, whose diagonal is $(0, 1, \ldots, p-1)$.

By definition, for each $\phi \in \mathcal{C}'$, $A_\phi \in \mathcal{L}_p$. The converse is also true. i.e., $\phi \to A_\phi$ is a one-to-one correspondence between $\mathcal{C}'$ and $\mathcal{L}_p$.

For $A \in \mathcal{L}_p$, let A_i be the ith row of A, viewed as a permutation of $\{0, 1, \ldots, p-1\}$, and let $\text{sign}(A) = \prod_{i=0}^{p-1} \text{sign}(A_i)$. Then $\text{sign}_D(\phi) = \text{sign}(A_\phi)$ and hence

$$\sum_{\phi \in \mathcal{C}'} sign_D(\phi) = \sum_{A \in \mathcal{L}_p} \text{sign}(A).$$

So our task is to show that

$$\sum_{A \in \mathcal{L}_p} \text{sign}(A) \neq 0. \tag{4.15}$$

For this purpose, we prove a stronger result:

$$\sum_{A \in \mathcal{L}_p} \text{sign}(A) \not\equiv 0 \pmod{p}. \tag{4.16}$$

This is a crucial step. By changing to inequality 4.16, our calculations will be carried out modulo p. This considerably reduces the computations. On the other hand, this step restricts our result to be valid only for the case when p is a prime number. We shall keep in mind that in the remainder of this section, all the calculations are carried out in the field $GF(p)$, i.e., integer calculation modulo p.

For a permutation π of $\{0, 1, \ldots, p-1\}$, and $A \in \mathcal{L}_p$, let $\pi(A)$ be the matrix obtained from A by permuting the rows, the columns and the symbols simultaneously. i.e., $\pi(A)$ is obtained from A by moving the ith row to the $\pi(i)$th row, the jth column to the $\pi(j)$th column, and then change each symbol i to $\pi(i)$. In other words,

$$(\pi(A))_{i,j} = \pi(A_{\pi^{-1}(i), \pi^{-1}(j)}).$$

For example, if $p = 3$, $\pi = (012)$ (i.e., $\pi(i) = i + 1$) and

$$A = \begin{pmatrix} 0 & 2 & 1 \\ 2 & 1 & 0 \\ 1 & 0 & 2 \end{pmatrix},$$

then to obtain $\pi(A)$, we 'move' row i to row $i + 1$, column i to column $i+1$, and then add 1 to each entry. In particular, $\pi(A)_{0,0} = A_{2,2} + 1 = 2 + 1 = 0$, $\pi(A)_{1,0} = A_{0,2} + 1 = 1 + 1 = 2$, $\pi(A)_{2,0} = A_{1,2} + 1 = 0 + 1 = 1$. Indeed, $\pi(A) = A$.

As another example, if $p = 5$ and $\pi = (01)$ (i.e., $\pi(0) = 1$, $\pi(1) = 0$ and $\pi(i) = i$ for $i = 2, 3, 4$) and

$$A = \begin{pmatrix} 0 & 3 & 1 & 4 & 2 \\ 3 & 1 & 4 & 2 & 0 \\ 1 & 4 & 2 & 0 & 3 \\ 4 & 2 & 0 & 3 & 1 \\ 2 & 0 & 3 & 1 & 4 \end{pmatrix},$$

then

$$\pi(A) = \begin{pmatrix} 0 & 3 & 4 & 2 & 1 \\ 3 & 1 & 0 & 4 & 2 \\ 4 & 0 & 2 & 1 & 3 \\ 2 & 4 & 1 & 3 & 0 \\ 1 & 2 & 3 & 0 & 4 \end{pmatrix}.$$

Lemma 4.9.4 *Assume that π is a permutation of $GF(p)$ and $A \in \mathcal{L}_p$. Then $\pi(A) \in \mathcal{L}_p$. Moreover, $\mathrm{sign}(A) = \mathrm{sign}(\pi(A))$.*

Proof. Assume that $A \in \mathcal{L}_p$. To show that $\pi(A) \in \mathcal{L}_p$, we need to verify Properties (1),(2),(3) in Definition 4.9.3 for $\pi(A)$. The verification is straightforward.

For the most part, it suffices to consider the case when $\pi = (ab)$ is an involution. Note that interchanging two rows does not change the sign of the matrix. Interchanging two columns of a row changes the parity of the corresponding permutation, and then interchanging two symbols in a row again changes the parity of the corresponding permutation. Therefore to obtain $\pi(A)$ from A, the sign is changed an even number of times, and hence $\mathrm{sign}(\pi(A)) = \mathrm{sign}(A)$. ∎

Let Γ be the cyclic group of permutations $\langle (01 \ldots (p-1)) \rangle$ generated by $(01 \ldots (p-1))$. Then Γ has order p. Consider the action of Γ on $\mathcal{L}_p$. As p is a prime, each orbit of the action is of size either p or 1.

Theorem 4.9.1 follows from the next lemma.

Lemma 4.9.5 *If p is an odd prime, then $\sum_{A \in \mathcal{L}_p} \text{sign}(A) \not\equiv 0$* (mod p).

The remainder of this section is devoted to the proof of Lemma 4.9.5.

The matrices in $\mathcal{L}_p$ are partitioned into orbits of Γ. By Lemma 4.9.4, each orbit of size p contributes a multiple of p to $\sum_{A \in \mathcal{L}_p} \text{sign}(A)$. Let $\mathcal{L}_p'$ be those matrices A that are fixed by Γ. To prove Lemma 4.9.5, it suffices to show that

$$\sum_{A \in \mathcal{L}_p'} \text{sign}(A) \not\equiv 0 \quad (\text{mod } p). \tag{4.17}$$

For a permutation $\mathbf{x} = (x_0, x_1, \ldots, x_{p-1})$ of $\{0, 1, \ldots, p-1\}$, let

$$A(\mathbf{x}) = \begin{pmatrix} x_0 & x_1 & x_2 & \cdots & x_{p-1} \\ x_{p-1}+1 & x_0+1 & x_1+1 & \cdots & x_{p-2}+1 \\ x_{p-2}+2 & x_{p-1}+2 & x_0+2 & \cdots & x_{p-3}+2 \\ \vdots & \vdots & \vdots & \cdots & \vdots \\ x_1+p-1 & x_2+p-1 & x_3+p-1 & \cdots & x_0+p-1 \end{pmatrix}.$$

Lemma 4.9.6 *For a permutation $\mathbf{x} = (x_0, x_1, \ldots, x_{p-1})$ of $\{0, 1, \ldots, p-1\}$, $A(\mathbf{x}) \in \mathcal{L}_p$ if and only if $x_0 = 0$ and $x_{p-i} = x_i - i$ for $i = 1, 2, \ldots, (p-1)/2$.*

Proof. The condition that $x_0 = 0$ is equivalent to the condition that the diagonal of $A(\mathbf{x})$ is $(0, 1, \ldots, p-1)$. The condition that $x_{p-i} = x_i - i$ for $i = 1, 2, \ldots, (p-1)/2$ is equivalent to saying that $A(\mathbf{x})$ is symmetric. ∎

Let $h = (p-1)/2$. For $\mathbf{x} = (x_1, x_2, \ldots, x_h) \in GF(p)^h$, let

$$\tau(\mathbf{x}) := (0, x_1, x_2, \ldots, x_h, x_h - h, \ldots, x_2 - 2, x_1 - 1),$$

$$\theta(\mathbf{x}) := (x_1, x_2, \ldots, x_h, x_h - h, \ldots, x_2 - 2, x_1 - 1).$$

Let

$$\mathcal{I}_p = \{\mathbf{x} \in GF(p)^h : \tau(\mathbf{x}) \text{ is a permutation of } \{0, 1, 2, \ldots, p-1\}\},$$
$$\theta(\mathcal{I}_P) = \{\theta(\mathbf{x}) : \mathbf{x} \in \mathcal{I}_p\}.$$

Lemma 4.9.7 *For $A \in \mathcal{L}_p$, A is fixed by Γ if and only if $A = A(\tau(\mathbf{x}))$ for some $\mathbf{x} \in \mathcal{I}_p$.*

Proof. Let $\pi = (01 \ldots (p-1))$. Assume that $A \in \mathcal{L}_p$ and $\pi(A) = A$. Assume that for $j = 0, 1, \ldots, p - 1$, $A_{0,j} = x_j$. By definition, $\pi(A)_{i,j} = A_{i-1,j-1} + 1$. Since $A_{i,j} = \pi(A)_{i,j}$, we conclude that $A_{i-1,j-1} = A_{i,j} - 1$. Apply this formula inductively, we have $A_{i,j} = x_{p-i+j} + i$, and hence $A = A(x_0, x_1, \ldots, x_{p-1})$. Since $A \in \mathcal{L}_p$, by Lemma 4.9.6, $(x_0, x_1, \ldots, x_{p-1}) = \tau(x_1, x_2, \ldots, x_h)$ and $(x_1, x_2, \ldots, x_h) \in \mathcal{I}_p$.

Conversely, if $(x_1, x_2, \ldots, x_h) \in \mathcal{I}_p$, then $A = A(\tau(\mathbf{x})) \in \mathcal{L}_p$ and $\pi(A) = A$ and hence A is fixed by Γ. ∎

Lemma 4.9.8 *For any permutation* $\mathbf{x} = (x_0, x_1, \ldots, x_{p-1})$ *of* $\{0, 1, \ldots, p - 1\}$, *all the rows of* $A(\mathbf{x})$ *have the same sign, and hence* $\operatorname{sign}(A(\mathbf{x})) = \operatorname{sign}(\mathbf{x})$.

Proof. (1) If $\mathbf{x} = (x_0, x_1, \ldots, x_{p-1})$ and $\mathbf{x}' = (x_{p-1}, x_0, \ldots, x_{p-2})$, then exactly $p - 1$ pairs (x_{p-1}, x_j) $(j \neq p - 1)$ changed status of being an inversion or not. Hence $\operatorname{sign}(\mathbf{x}) = \operatorname{sign}(\mathbf{x}')$.

(2) Assume $\mathbf{x} = (x_0, x_1, \ldots, x_{p-1})$ and $\mathbf{x}' = (x_0 + 1, x_1 + 1, \ldots, x_{p-1} + 1)$, where additions are modulo p. Assume $x_{i^*} = p - 1$. Then $x_{i^*} + 1 = 0$, and exactly $p - 1$ pairs (x_{i^*}, x_j) $(j \neq i^*)$ changed status of being an inversion or not. Hence $\operatorname{sign}(\mathbf{x}) = \operatorname{sign}(\mathbf{x}')$.

All the rows of $A(\mathbf{x})$ are obtained from the first row by repeatedly applying the above two operations. Hence all the rows of $A(\mathbf{x})$ have the same sign. As A has an odd number of rows, $\operatorname{sign}(A(\mathbf{x})) = \operatorname{sign}(\mathbf{x})$. ∎

Note that $\operatorname{sign}(\tau(\mathbf{x})) = \operatorname{sign}(\theta(\mathbf{x}))$. Hence we have the following corollary.

Corollary 4.9.9 *For any odd prime* p, $\sum_{A \in \mathcal{L}_p} \operatorname{sign}(A) = \sum_{\mathbf{x} \in \mathcal{I}_p} \operatorname{sign}(\tau(\mathbf{x})) = \sum_{\mathbf{x} \in \mathcal{I}_p} \operatorname{sign}(\theta(\mathbf{x}))$. *Here the calculation is in* $GF(p)$, *i.e., the summation is modulo* p.

It remains to show that (in $GF(p)$)

$$\sum_{\mathbf{x} \in \theta(\mathcal{I}_P)} \operatorname{sign}(\mathbf{x}) \neq 0. \tag{4.18}$$

For the purpose of proving inequality 4.18, we introduce another polynomial $f(\mathbf{x}) \in GF(p)[x_1, x_2, \ldots, x_h]$ of degree $h(p - 1)$ so that 4.18 holds if and only if for $\mathbf{t} = (p - 1, p - 1, \ldots, p - 1)$, $c_{f,\mathbf{t}} \neq 0$.

For $\mathbf{x} = (x_1, x_2, \ldots, x_{p-1}) \in GF(p)^{p-1}$, let

$$\phi(\mathbf{x}) := \left(\prod_{1 \leq i < j \leq p-1, j \neq p-i} (x_j - x_i) \right) \left(\prod_{i=1}^{p-1} x_i \right).$$

For $\mathbf{x} = (x_1, x_2, \ldots, x_h) \in GF(p)^h$, let

$$f(\mathbf{x}) := \phi(\theta(\mathbf{x})).$$

Then $\phi(\mathbf{x}) \in GF(p)[x_1, x_2, \ldots, x_{p-1}]$ is a polynomial of degree $deg(\phi) = \frac{p(p-1)}{2} = h(p-1)$ with $p-1$ variables, and $f(\mathbf{x}) \in GF(p)[x_1, x_2, \ldots, x_h]$ is also a polynomial of degree $h(p-1)$, but with $h = (p-1)/2$ variables.

For $\mathbf{x} \in GF(p)^{p-1}$, $\phi(\mathbf{x}) \neq 0$ implies that the x_i's are nonzero and they are pairwise distinct, except possibly $x_i = x_{p-i}$ for some i. If $\mathbf{x} \in \theta(\mathcal{I}_P)$, then we know that $x_i - x_{p-i} = i \neq 0$ for all $i = 1, 2, \ldots, h$. Hence for $\mathbf{x} \in GF(p)^h$, $f(\mathbf{x}) \neq 0$ if and only if $\theta(\mathbf{x})$ is a permutation of $\{1, 2, \ldots, p-1\}$, which means that $\mathbf{x} \in \mathcal{I}_p$.

Let $\mathbf{t} = (p-1, p-1, \ldots, p-1) \in \mathbb{N}_0^h$. By Corollary 4.4.2, the coefficient $c_{f,\mathbf{t}}$ of the monomial $\mathbf{x}^{\mathbf{t}} = \prod_{i=1}^h x_i^{p-1}$ in the expansion of $f(\mathbf{x})$ is equal to

$$c_{f,\mathbf{t}} = (-1)^h \sum_{\mathbf{x} \in GF(p)^h} f(\mathbf{x}).$$

Let $\mathbf{x}_0 \in GF(p)^h$ be the vector such that $\theta(\mathbf{x}_0) = (1, 2, \ldots, p-1)$. Then $\text{sign}(\theta(\mathbf{x})) = \frac{f(\mathbf{x})}{f(\mathbf{x}_0)}$. Since $f(\mathbf{x}_0) \neq 0$ (in $GF(p)$), to show that $\sum_{\mathbf{x} \in \theta(\mathcal{I}_p)} \text{sign}(\mathbf{x}) \neq 0$ (in $GF(p)$), it suffices to show that

$$c_{f,\mathbf{t}} = (-1)^h \sum_{\mathbf{x} \in GF(p)^h} f(\mathbf{x}) \neq 0. \tag{4.19}$$

For the purpose of proving inequality 4.19, we consider a third polynomial g, such that the the monomial $\prod_{i=1}^h x_i^{p-1}$ has the same coefficient in the expansion of $g(\mathbf{x})$ as that in $f(\mathbf{x})$, and there is a simple formula to calculate this coefficient by applying Corollary 4.4.3 again. This is a useful technique in the applications of Corollary 4.4.3, and is used in several papers for different problems.

The polynomial $f(\mathbf{x})$ is the product of terms of the form $\pm(x_j - x_i - a)$ for $1 \leq i < j \leq h$ (where a is some constant) and terms of the form $(x_i - a)$ for $1 \leq i \leq h$. For each pair $1 \leq i < j \leq h$, there are four terms of the form $\pm(x_j - x_i - a)$ and for each $1 \le i \le h$, there are two terms of the form $(x_i - a)$. Changing the constants a in the terms of the product will not change the coefficient of the highest degree monomial in the expansion.

Let

$$g(\mathbf{x}) := \left(\prod_{1 \le i < j \le h} (x_j - x_i)^2 \right) \left(\prod_{1 \le i < j \le h} ((x_j - x_i)^2 - 1) \right)$$
$$\times \left(\prod_{i=1}^{h} x_i(x_i - 1) \right).$$

Note that $(x_j - x_i)^2 - 1 = (x_j - x_i - 1)(x_j - x_i + 1)$. So $g(\mathbf{x})$ is obtained from $f(\mathbf{x})$ by replacing each term $\pm(x_j - x_i - a)$ with a term $(x_j - x_i - a')$ for some constant a' which might be distinct from a, and replacing each term $(x_i - a)$ with a term $(x_i - a'')$ for some constant a'' which might be distinct from a. Therefore, the coefficients of the highest degree monomials in $g(\mathbf{x})$ and $f(\mathbf{x})$ differ by at most a sign $\pm$. In particular,

$$c_{g,\mathbf{t}} = \pm c_{f,\mathbf{t}}.$$

By Corollary 4.4.2,

$$c_{g,\mathbf{t}} = (-1)^h \sum_{\mathbf{x} \in \mathbb{F}^h} g(\mathbf{x}).$$

Now for $\mathbf{x} \in GF(p)^h$, $g(\mathbf{x}) \ne 0$ if and only if $\mathbf{x}$ is a permutation of $\{2, 4, \ldots, p-1\}$. Indeed, $\prod_{i=1}^{h} x_i(x_i - 1) \ne 0$ implies that $x_i \notin \{0, 1\}$; $\prod_{1 \le i < j \le h} (x_i - x_j)^2 \ne 0$ implies that $x_1, x_2, \ldots, x_h$ are pairwise distinct; and $\prod_{1 \le i < j \le h} ((x_i - x_j)^2 - 1) \ne 0$ implies that $x_1, x_2, \ldots, x_h$ are pairwise non-consecutive. Now for permutations $\mathbf{x}$ of $\{2, 4, \ldots, p-1\}$, $g(\mathbf{x})$ is a non-zero constant C'. So

$$c_{g,\mathbf{t}} = (-1)^{-h} C' h! \ne 0.$$

This completes the proof of Theorem 4.9.1. ∎

4.10   Jaeger's conjecture

The following conjecture concerning bases and hyperplanes was raised by Jaeger [30] (originally stated for $q = 5$ only).

Conjecture 4.10.1 *To any two bases B and B' of an n-dimensional vector space over a finite field $GF(q)$ with $q > 3$, there exists a hyperplane H that is disjoint from both B and B'.*

Using the vectors in B as row vectors, we form an $n \times n$ matrix A. The statement that a hyperplane $\mathbf{a}^\perp$ is disjoint from B is equivalent to the statement that the vector $A\mathbf{a}$ has no zero entry, i.e., $(A\mathbf{a})_j \neq 0$ for $j = 1, 2, \ldots, n$. By a change of coordinates, if needed, we may assume that A is the identity matrix. Then the requirement that $A\mathbf{a}$ has no zero entry means that $\mathbf{a}$ has no zero entry. This motivates the following definition.

Definition 4.10.2 *Assume that $\mathbb{F}$ is a field, $\mathbf{x} \in \mathbb{F}^n$ is a vector and $A \in \mathbb{F}^{m \times n}$ is an $m \times n$ matrix. If neither $\mathbf{x}$ nor $A\mathbf{x}$ has a zero entry, i.e., $x_i \neq 0$ for $i = 1, 2, \ldots, n$ and $(A\mathbf{x})_j \neq 0$ for $j = 1, 2, \ldots, m$, then $\mathbf{x}$ is called a* nowhere zero point *of A.*

By this definition, Jaeger's conjecture is equivalent to the following conjecture, which was formulated by Alon and Tarsi [2].

Conjecture 4.10.3 *Every non-singular $n \times n$ matrix A, over a finite field $GF(q)$ with $q > 3$ elements, has a nowhere zero point.*

Consider the matrix

$$A = \begin{bmatrix} 1 & 1 \\ 1 & -1 \end{bmatrix}$$

in $\mathbb{F}_3^{2 \times 2}$. Note that A is a non-singular matrix. If $x_1, x_2 \in \mathbb{F}_3 - \{0\} = \{1, -1\}$, then either $x_1 + x_2 = 0$ or $x_1 - x_2 = 0$. This means that A does not have a nowhere zero point.

So the condition that $q > 3$ is needed in the conjecture.

Alon and Tarsi proved the following theorem, which confirms the conjecture for the case when $q = p^\alpha$ where p is a prime and $\alpha \geq 2$ is an integer. The proof of this theorem presented below is a different proof given by Schauz and Honold [54]. There are two useful techniques used in the proof, which are also used in a few other applications of CNS.

(1) To calculate the coefficient of a monomial $m(x)$ in a polynomial f, we may apply Theorem 4.3.1 to another polynomial $\tilde{f}$, provided that $m(x)$ has the same coefficient in both f and $\tilde{f}$. By choosing an appropriate $\tilde{f}$, the formula in Theorem 4.3.1 can be made simpler and easier to compute.

(2) The grid $\mathbf{S}$ used in the calculation of the coefficient of a monomial in f can be different from the grid $\mathbf{T}$ on which the CNS is applied to find a non-zero point, provided that $\mathbf{S}$ and $\mathbf{T}$ have the same size. Again by choosing an appropriate grid $\mathbf{S}$, the calculation becomes much easier.

Theorem 4.10.4 *Assume that q is a proper prime power, i.e., $q = p^\alpha$ with p being a prime and $\alpha \geq 2$. Then every non-singular $n \times n$ matrix A over $GF(q)$ has a nowhere zero point.*

Proof. As per our assumption, $q = p^\alpha$, where p is a prime and $\alpha \geq 2$, and A is an $n \times n$ non-singular matrix over $GF(q)$.

For $\mathbf{x} \in GF(q)^n$, denote by $(A\mathbf{x})_j$ the jth component of $A\mathbf{x}$. Let

$$P(\mathbf{x}) = \prod_{j=1}^{n}(A\mathbf{x})_j^{p-1}.$$

Then $P(\mathbf{x}) \neq 0$ if and only if $(A\mathbf{x})_j \neq 0$ for each j.

Observe that the degree of $P(\mathbf{x})$ is $n(p-1)$. Let $\mathbf{t} = (p-1, p-1, \ldots, p-1)$. If $c_{P,\mathbf{t}} \neq 0$, then for any grid $\mathbf{S}$ with $||\mathbf{S}|| = \mathbf{t}$, there exists $\mathbf{a} \in \mathbf{S}$ for which $P(\mathbf{a}) \neq 0$. The condition that $\alpha \geq 2$ implies that $GF(q)$ has a set S consisting of p non-zero elements (this is the only place that we use the assumption that $\alpha \geq 2$). Let $\mathbf{S} = S^n$. Then a vector $\mathbf{a} \in \mathbf{S}$ with $A\mathbf{a} \neq 0$ would be a nowhere zero point of A. So to prove Theorem 4.10.4, it suffices to show that $c_{P,\mathbf{t}} \neq 0$. Let

$$\tilde{P}(\mathbf{x}) = \prod_{j=1}^{n}((A\mathbf{x})_j^{p-1} - 1).$$

It follows from the definition that $c_{P,\mathbf{t}} = c_{\tilde{P},\mathbf{t}}$. So it is enough to show that $c_{\tilde{P},\mathbf{t}} \neq 0$. We shall use Theorem 4.3.1 to prove this.

Claim 4.10.5 *Let $\mathbf{S} := \{0, 1, \ldots, p-1\}^n$. Then*

$$\sum_{\mathbf{b} \in \mathbf{S}} N_{\mathbf{S}}^{-1}(\mathbf{b})\tilde{P}(\mathbf{b}) = \det(A)^{p-1}.$$

Proof of Claim 5 By (3) of Lemma 4.4.1, for $\mathbf{b} \in \mathbf{S}$, $N_{\mathbf{S}}^{-1}(\mathbf{b}) = (-1)^n$. Assume that $A = [a_{ij}]_{n \times n}$. Let

$$
\begin{aligned}
f(a_{1,1}, a_{1,2}, \ldots, a_{n,n}) &= \sum_{\mathbf{b} \in \mathbf{S}} N_{\mathbf{S}}^{-1}(\mathbf{b}) \tilde{P}(\mathbf{b}) - \det(A)^{p-1} \\
&= \sum_{\mathbf{b} \in \mathbf{S}} (-1)^n \tilde{P}(\mathbf{b}) - \det(A)^{p-1}.
\end{aligned}
$$

Then f is a polynomial in the n^2 variables $a_{i,j}$ ($1 \le i \le n, 1 \le j \le n$) of degree $n(p-1)$, and in which each variable $a_{i,j}$ has highest degree at most $p-1$. We shall prove that f is the constant zero polynomial. Otherwise, by CNS, f has a non-zero point in the grid $\{0, 1, \ldots, p-1\}^{n^2}$.

Assume that $(a_{1,1}, \ldots, a_{n,n}) \in \{0, 1, \ldots, p-1\}^{n^2}$ is a non-zero point of f. Then $A = [a_{i,j}]$ is an $n \times n$ matrix whose entries lie in the prime field $GF(p)$. For $\mathbf{b} \in \mathbf{S}$,

$$
\tilde{P}(\mathbf{b}) = \begin{cases} (-1)^n, & \text{if } A\mathbf{b} = \mathbf{0}, \\ 0, & \text{otherwise.} \end{cases}
$$

Thus

$$
\sum_{\mathbf{b} \in \mathbf{S}} (-1)^n \tilde{P}(\mathbf{b}) = |ker(A)|1 = \begin{cases} 1, & \text{if } A \text{ is non-singular}, \\ 0, & \text{otherwise.} \end{cases}
$$

Here $ker(A)$ is the kernel of A, i.e., the set of solutions of $A\mathbf{x} = \mathbf{0}$ in $GF(p)$. If A is non-singular, then $ker(A)$ consists of a single vector, i.e., the $\mathbf{0}$ vector. If A is singular, then $|ker(A)|$ is a multiple of p and hence $|ker(A)|1 = 0$ in $GF(p)$. As

$$
\det(A)^{p-1} = \begin{cases} 1, & \text{if } A \text{ is non-singular}, \\ 0, & \text{otherwise}, \end{cases}
$$

we have $f(a_{1,1}, a_{1,2}, \ldots, a_{n,n}) = 0$. This completes the proof of Claim 4.10.5. ∎

It follows from Claim 4.10.5 and Theorem 4.3.1 that if A is an $n \times n$ non-singular matrix over $GF(q)$, then

$$
c_{\tilde{P}, \|\mathbf{S}\|} = \sum_{\mathbf{a} \in \mathbf{S}} N_{\mathbf{S}}(\mathbf{a})^{-1} \tilde{P}(\mathbf{a}) = \det(A)^{p-1} \ne 0.
$$

This completes the proof of Theorem 4.10.4. ∎

Chapter 5

Permanent and Vertex-edge Weighting

Let $A = [a_{ij}]$ be a square matrix of order n. The *permanent of A* is defined as

$$\text{per}(A) = \sum_{\sigma \in S_n} \prod_{i=1}^{n} a_{i\sigma(i)}.$$

In this chapter, we express the coefficient of monomials of a polynomial $f(\mathbf{x})$ as the permanent of certain matrices and use this formula to study vertex-edge weightings of graphs.

5.1 Permanent as the coefficient

Assume that $f(\mathbf{x}) \in \mathbb{F}[x_1, x_2, \ldots, x_n]$ is defined as the product of linear polynomials

$$f(\mathbf{x}) = \prod_{i=1}^{m} \sum_{j=1}^{n} a_{ij} x_j.$$

Then the coefficient of the monomial $\mathbf{x}^{\mathbf{t}}$ in the expansion of f can be calculated as the permanent of a certain matrix.

Let $A = [a_{ij}]_{m \times n}$ be the coefficient matrix of f. We denote by A_j the jth column of A. For a mapping $\eta : \{1, 2, \ldots, n\} \to \{0, 1, \ldots\}$, let $A(\eta)$ be the matrix such that each column of $A(\eta)$ is a column of A, and for $j \in \{1, 2, \ldots, n\}$, $A(\eta)$ contains $\eta(j)$ copies of A_j. We prove that the coefficient of the monomial $\mathbf{x}^{\eta} = \prod_{i=1}^{n} x_i^{\eta(i)}$ equals $\frac{1}{\prod_{i=1}^{n} \eta(i)!} \text{per}(A(\eta))$.

For example, consider the polynomial

$$f(\mathbf{x}) = (x_1 + 2x_2 + 3x_3)(2x_1 + 3x_2 + 4x_3)(x_2 + 2x_3)(2x_1 + x_2 + 2x_3).$$

To expand f, we first select one entry from each of the four factors $x_1 + 2x_2 + 3x_3, 2x_1 + 3x_2 + 4x_3, 0x_1 + x_2 + 2x_3$ and $2x_1 + x_2 + 2x_3$, multiply these entries to get a monomial. There are 3^4 possible selections; each can be expressed as a mapping $\tau : \{1, 2, 3, 4\} \rightarrow \{1, 2, 3\}$, where $\tau(i) = j$ means that we select the term x_j in the ith factor. Then take the summations over all such selections to obtain the expansion of f. When calculating the coefficient of a particular monomial in the expansion of f, we concentrate on those selections that contribute to its coefficient. For example, to calculate the coefficient of monomial $x_1^2 x_3^2$, we need to choose x_1 twice and x_3 twice in the product. The coefficient matrix of f is

$$A = \begin{pmatrix} 1 & 2 & 3 \\ 2 & 3 & 4 \\ 0 & 1 & 2 \\ 2 & 1 & 2 \end{pmatrix}$$

When multiplying out the polynomial f, we select one column from each of the 4 rows. For the product to contribute to the monomial $x_1^2 x_3^2$, we need to select twice from the 1st column and twice from the 3rd column. Such a selection can be associated to the number of permutations of $\{1, 2, 3, 4\}$, which are in one-to-one correspondences with the columns and rows of $A(\eta)$, where $\eta(1) = \eta(3) = 2$ and $\eta(2) = 0$. Indeed, there are $\binom{4}{2} = 6$ selections, and there are $4!$ permutations of $\{1, 2, 3, 4\}$. Each of the $\binom{4}{2}$ selections corresponds to $2!2!$ permutations of $\{1, 2, 3, 4\}$. For example, $\tau(1) = \tau(2) = 1, \tau(3) = \tau(4) = 3$ is such a selection. For this selection, the multiplication of the selected entries gives $x_1 \times 2x_1 \times 2x_3 \times 2x_3 = 8x_1^2 x_3^2$. The matrix $A(\eta)$ is

$$A(\eta) = \begin{pmatrix} 1 & 1 & 3 & 3 \\ 2 & 2 & 4 & 4 \\ 0 & 0 & 2 & 2 \\ 2 & 2 & 2 & 2 \end{pmatrix}.$$

The above selection function τ corresponds to 4 permutations in the calculation of the permanent of $A(\eta)$, namely $id, (12), (34), (12)(34)$. Apply this argument to general polynomials f of the form $f(x_1, x_2, \ldots, x_n) = \prod_{i=1}^{m} \sum_{j=1}^{n} a_{ij} x_j$, we have the following lemma.

Lemma 5.1.1 *Assume that $f(x_1, x_2, \ldots, x_n)$ and $A(\eta)$ are defined as above. Then the coefficient of the monomial $\mathbf{x}^\eta$ in the expansion of f equals $\frac{1}{\prod_{i=1}^{n} \eta(i)!} \mathrm{per}(A(\eta))$.*

Proof. Let Γ be the set of all mappings $\tau : \{1, 2, \ldots, m\} \to \{1, 2, \ldots, n\}$, and each $\tau \in \Gamma$ is called a selection mapping. Then

$$f(x_1, x_2, \ldots, x_n) = \sum_{\tau \in \Gamma} \prod_{i=1}^{m} a_{i\tau(i)} x_{\tau(i)}.$$

For $\tau \in \Gamma$, let $\mathbf{d}(\tau) = (d_1, d_2, \ldots, d_n)$, where $d_i = |\{j : \tau(j) = i\}|$. Then the monomial corresponding to the selection function $\tau \in \Gamma$ is $\mathbf{x}^{\mathbf{d}(\tau)}$. The coefficient of the monomial $\mathbf{x}^{\eta}$ in the expansion of $f(\mathbf{x})$ equals

$$\sum_{\mathbf{d}(\tau) = (\eta(1), \eta(2), \ldots, \eta(n))} \prod_{i=1}^{m} a_{i\tau(i)}.$$

In other words, to make a contribution to the coefficient of $\mathbf{x}^{\eta}$, we consider those selection functions τ that choose $\eta(i)$ times from the ith column of A.

Now instead of choosing $\eta(i)$ times from the ith column of A, we make the matrix $A(\eta)$ by duplicating $\eta(i)$ copies of the ith column of A. Then the selection functions τ described above correspond to selecting once in each column of $A(\eta)$. For a fixed selection function τ, there are $\prod_{i=1}^{n} \eta(i)!$ permutations π such that $\tau(j) = i$ if and only if $\pi(j)$ is the index of a column of $A(\eta)$ as a copy of A_i. Thus each selection function $\tau \in \Gamma$ corresponds to $\prod_{i=1}^{n} \eta(i)!$ permutations in the calculation of $\mathrm{per}(A(\eta))$. Hence the coefficient of the monomial $\mathbf{x}^{\eta}$ in the expansion of f equals $\frac{1}{\prod_{i=1}^{n} \eta(i)!} \mathrm{per}(A(\eta))$. ∎

By Lemma 5.1.1, to prove that the monomial $\mathbf{x}^{\eta}$ of f is non-vanishing, it amounts to proving that $A(\eta)$ has non-zero permanent.

The graph polynomial $f_G(\mathbf{x})$ is defined as the product of linear polynomials, where the coefficent matrix is the incidence matrix B_D of an orientation D of G. To be precise, if G has n vertices $v_1, v_2, \ldots, v_n$ and m edges $e_1, e_2, \ldots, e_m$, and D is an orientation of G, then B_D is an $m \times n$ matrix in which

$$(B_D)_{i,j} = \begin{cases} 1, & \text{if } v_j \text{ is the tail of } e_i, \\ -1, & \text{if } v_j \text{ is the head of } e_i, \\ 0, & \text{otherwise,} \end{cases}$$

It follows from the discussion above that for any graph G, $AT(G) = \min\{k \colon \mathrm{per}(B_D(\eta)) \neq 0 \text{ for some } \eta : E(G) \to \{0, 1, \ldots, k-1\}\}$.

Although the definition of the permanent of a matrix is very similar to the definition of determinant of a matrix, the calculation of the permanent of a matrix is a much harder problem. Using the Gaussian Elimination Method, one can calculate the determinant of a matrix in polynomial time. However, it is NP-hard to determine the permanent of a matrix. Nevertheless, the permanent method is useful in some combinatorial problems. In this chapter, we apply this method to the study of vertex-edge weightings of graphs.

The following property of permanents will be frequently used. It follows easily from the definition and the proof is therefore omitted.

Lemma 5.1.2 *Assume that* $C = [A_1 + B_1, A_2, \ldots, A_m]$ *is a square matrix of order* m *with* A_i, B_1 *being column vectors of length* m. *Let* $A = [A_1, A_2, \ldots, A_m]$ *and* $B = [B_1, A_2, \ldots, A_m]$, *then*

$$per(C) = per(A) + per(B).$$

5.2   Edge weighting and total weighting

Definition 5.2.1 *Assume that* $\mathbb{F}$ *is a field. A* total weighting *over* $\mathbb{F}$ *of a graph* G *is a mapping* $\phi : V(G) \cup E(G) \to \mathbb{F}$. *A total weighting* ϕ *is* proper *if for any edge* uv *of* G,

$$\sum_{e \in E(u)} \phi(e) + \phi(u) \neq \sum_{e \in E(v)} \phi(e) + \phi(v).$$

A total weighting with $\phi(v) = 0$ *for each* $v \in V$ *is called an* edge-weighting *of* G. *A proper total weighting with* $\phi(v) = 0$ *for each* $v \in V$ *is called a* proper edge-weighting *of* G.

Definition 5.2.2 *For a subset* S *of* $\mathbb{F}$, *a proper total weighting* ϕ *of* G *with* $\phi(z) \in S$ *for all* $z \in V(G) \cup E(G)$ *is called an* S-total weight colouring *of* G. *A proper edge weighting* ϕ *of* G *with* $\phi(e) \in S$ *for all* $e \in E(G)$ *is called an* S-edge weight colouring *of* G.

For the study of edge weighting and total weighting, if not stated explicitly otherwise, the field is assumed to be $\mathbb{R}$. Figure 5.1 is a $\{1, 2, 3\}$-edge weight colouring of $G = C_5$. The weight of v_1 equals the sum of the weights of the edges incident at v_1, which is $1 + 2 = 3$.

The 1-2-3 conjecture of Karonski, Luczak and Thomason [33] is the following.

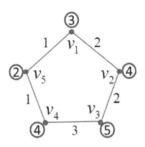

FIGURE 5.1: A $\{1, 2, 3\}$ – edge weighting of $G = C_5$

Conjecture 5.2.3 [1-2-3 conjecture] *Every graph containing no isolated edge has a $\{1, 2, 3\}$-edge weight colouring.*

The 1–2 conjecture of Przybylo and Wozniak [50] states the following:

Conjecture 5.2.4 [1-2 conjecture] *Every graph G has a $\{1, 2\}$-total weight colouring.*

Both the conjectures have been verified for some special graph classes but remain open in general. The best known results concerning these two conjectures are the following results of Kalkowski, Karonski and Pfender [32] and of Kalkoswki [31].

Theorem 5.2.5 *[32] Every graph with no isolated edges is $\{1, 2, 3, 4, 5\}$-edge weight colourable.*

Theorem 5.2.6 *[31] Every graph has a proper total weighting ϕ with $\phi(v) \in \{1, 2\}$ for every vertex v, and $\phi(e) \in \{1, 2, 3\}$ for every edge e.*

For the application of CNS to this problem, we are interested in the list version of edge weighting and total weighting of graphs. The list version of edge-weighting was first studied by Bartnicki, Grytczuk and Niwcyzk [10].

Definition 5.2.7 *Assume that $\mathbb{F}$ is a field. A k-edge list assignment of G is a mapping L which assigns to each edge e of G a subset $L(e)$ of $\mathbb{F}$ of size k. A proper L-edge weighting is a proper edge weighting ϕ with $\phi(e) \in L(e)$ for all $e \in E(G)$. A graph G is said to be k-edge weight choosable if for any k-edge list assignment L of G, there is a proper L-edge weighting of G.*

It follows from the definition that if G is k-edge weight choosable, then G is $\{1, 2, \ldots, k\}$-edge weight colourable. The converse is not true.

Figure 5.2(a) gives a $\{1, 2\}$Yes.-edge-weight colouring of the diamond graph.

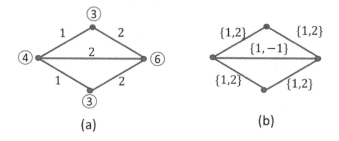

(a)   (b)

FIGURE 5.2: (a) A proper edge weighting; (b) an edge list assignment.

Figure 5.2 shows a graph which is $\{1, 2\}$-edge weight colourable, but not 2-edge weight choosable. For the 2-edge list assignment L in Figure 5.2(b), it is impossible to obtain a proper edge weighting ϕ of the diamond graph with $\phi(e) \in L(e)$ for every edge e. Nevertheless, Bartnicki, Grytczuk and Niwcyzk [10] proposed the following conjecture, which is stronger than Conjecture 5.2.3.

Conjecture 5.2.8 [3-edge weight choosable conjecture] *Every graph without an isolated edge is 3-edge-weight choosable.*

The list version of total weighting of graphs was introduced by Przybylo and Wozniak [51] snd independently by Wong and Zhu [64].

Definition 5.2.9 *Assume that $\mathbb{F}$ is a field and $\eta : V(G) \cup E(G) \to \{1, 2, \ldots, \}$ is a mapping which assigns to each $z \in V(G) \cup E(G)$ a positive integer $\eta(z)$. A total η-list assignment is a mapping L which assigns to each $z \in V(G) \cup E(G)$ a subset $L(z)$ of $\mathbb{F}$ of size $\eta(z)$. We say that G is said to be η-choosable if for any total η-list assignment L, there is a proper L-total weighting of G. If $\eta(v) = k$ for each vertex v and $\eta(e) = k'$ for each edge e, then a total η-list assignment is called a total (k, k')-list assignment of G. We say that G is (k, k')-choosable if for any (k, k')-list assignment L, there is a proper L-total weighting of G.*

Lemma 5.2.10 *A graph G is $(k, 1)$-choosable if and only if G is k-vertex choosable.*

Proof. The 'only if' is trivial. Now we prove the 'if' part. Assume that G is k-vertex choosable and L is a $(k, 1)$ list for G. Assume that $L(e) = \{w_e\}$ for each edge e of G. For each vertex v of G, let $c_v = \sum_{e \in E(v)} w_e$. Let $L'(v) = \{w - c_v : w \in L(v)\}$. Then L' is a k-list assignment for G. Hence G has an L'-colouring f. Let $g(v) = f(v) - c_v$ for each vertex v, and $g(e) = w_e$ for each edge e. Then g is a proper L-total weighting of G. ∎

Remark 5.2.11 G *is* $(1, k)$-*choosable* $\Rightarrow$ G *is* k-*edge-weight-choosable.*

The converse of Remark 5.2.11 is not true. The path P_4 is 2-edge weight choosable but it is not $(1, 2)$-choosable. (Hint: Take the vertex weights to be $1, 0, 0, 0$ in order and $L(e) = \{0, 1\}$ for each $e \in P_4$).

As strengthenings of Conjectures 5.2.3 and 5.2.4, the following two conjectures are proposed in [64]:

Conjecture 5.2.12 [$(1, 3)$-**choosable conjecture**] *Every graph with no isolated edges is* $(1, 3)$-*choosable.*

Conjecture 5.2.13 [$(2, 2)$-**choosable conjecture**] *Every graph is* $(2, 2)$-*choosable.*

Conjecture 5.2.13 is also proposed in [51].

5.3 Polynomial associated to total weighting

Assume that G is a graph whose vertices are linearly ordered. We associate to each vertex v a variable x_v and to each edge e a variable x_e. Let $\tilde{P}_G(\mathbf{x}) \in \mathbb{F}[x_z : z \in V \cup E]$ be the polynomial defined by

$$\tilde{P}_G(\mathbf{x}) = \prod_{v < u, uv \in E(G)} \left((x_v + \sum_{e \in E(v)} x_e) - (x_u + \sum_{e \in E(u)} x_e) \right).$$

For a total weighting ϕ of G, $\tilde{P}_G(\phi)$ is the evaluation of P_G with $x_v = \phi(v)$ and $x_e = \phi(e)$ for every $v \in V$ and $e \in E$. Then ϕ is a proper total weighting of G if and only if $\tilde{P}_G(\phi) \neq 0$.

The coefficient matrix, denoted by $A_G = [a_{ez}]$, of $\tilde{P}_G$ is the matrix whose rows are indexed by the edges of G, and columns are indexed by the vertices and edges of G, where for an edge $e = uv$ of G with $u < v$, and for $z \in V(G) \cup E(G)$,

$$a_{ez} = \begin{cases} 1, & \text{if either } z = u \text{ or } z \neq e \text{ is an edge of } G \text{ incident to } u, \\ -1, & \text{if either } z = v \text{ or } z \neq e \text{ is an edge of } G \text{ incident to } v, \\ 0, & \text{otherwise.} \end{cases}$$

Hence if G has n vertices and m edges, then A_G is an m by $(n + m)$ matrix.

A column of A_G indexed by a vertex v is denoted by $A_G(v)$ and is called a vertex column of A_G, and a column of A_G indexed by an edge e is denoted by $A_G(e)$ and is called an edge column of A_G. We denote by B_G the submatrix of A_G consisting of the edge columns of A_G.

For an edge $e = uv$ with $u < v$, we say that u is the *smaller endvertex* of e and v is the *larger endvertex* of e.

Lemma 5.3.1 *Assume that G is a graph and $e = uv \in E(G)$. Then*

$$A_G(e) = A_G(u) + A_G(v).$$

Proof. It suffices to show that for any edge e, $a_{e'e} = a_{e'u} + a_{e'v}$. This follows from the facts below:

$$a_{e'u} = \begin{cases} 1, & \text{if } u \text{ is the smaller endvertex of } e', \\ -1, & \text{if } u \text{ is the larger endvertex of } e', \\ 0, & \text{otherwise.} \end{cases}$$

$$a_{e'e} = \begin{cases} 1, \text{if } e \neq e' \text{ and } e \text{ is incident to} \\ \quad \text{the smaller endvertex of } e', \\ -1, \text{ if } e \neq e' \text{ and } e \text{ is incident to} \\ \quad \text{the larger endvertex of } e', \\ 0, \text{ otherwise.} \end{cases}$$

Therefore if e' is not incident to any of u or v, then $a_{e'e} = 0, a_{e'u} = a_{e'v} = 0$. If $e \neq e'$ and u or v is the smaller endvertex of e', then $a_{e'e} = 1$ and $a_{e'u} + a_{e'v} = 1$. If $e \neq e'$ and u or v is the larger endvertex of e', then $a_{e'e} = -1$ and $a_{e'u} + a_{e'v} = -1$. If $e = e'$ then $a_{e'e} = 0$, $a_{e'u} = 1, a_{e'v} = -1$ (assume that $u < v$), so $a_{e'u} + a_{e'v} = 0$. ∎

Lemma 5.3.2 *Assume that A is an $n \times m$ matrix and L is an $n \times n$ matrix whose columns are linear combinations of the columns of A. Let the jth column of A be present in n_j such linear combinations (with non-zero coefficients). Then there is an index function η : $\{1, 2, \ldots, m\} \to \{0, 1, \ldots\}$ such that $\eta(j) \leq n_j$ and $\mathrm{per}(A(\eta)) \neq 0$.*

Proof. For $1 \leq k \leq n$, let I_k be the set of indices j such that

$$L_k = \sum_{j \in I_k} \alpha_{k,j} A_j,$$

where $\alpha_{k,j} \neq 0$.

Let $\mathcal{F}$ be the family of mappings $f : \{1, 2, \ldots, n\} \to \{1, 2, \ldots, m\}$ such that $f(k) \in I_k$ for $k = 1, 2, \ldots, n$. For each $f \in \mathcal{F}$, let L_f be the matrix obtained from L by replacing the kth column L_k of L with $A_{f(k)}$. As the permanent of a matrix is a multilinear function with respect to its columns, we have

$$\mathrm{per}(L) = \sum_{f \in \mathcal{F}} \prod_{k=1}^{n} \alpha_{k,f(k)} \mathrm{per}(L_f).$$

Since $\mathrm{per}(L) \neq 0$, $\mathrm{per}(L_f) \neq 0$ for some $f \in \mathcal{F}$. For $j = 1, 2, \ldots, m$, let $\eta(j) = |f^{-1}(j)| \leq n_j$. Then $L_f = A(\eta)$, and hence $\mathrm{per}(A(\eta)) \neq 0$. ∎

The following is a corollary of Lemma 5.1.1 and CNS.

Corollary 5.3.3 *If there is a mapping $\eta : V \cup E \to \{0, 1, \ldots\}$ such that $\mathrm{per}(A_G(\eta)) \neq 0$, and $\eta'(z) = \eta(z) + 1$ for $z \in V(G) \cup E(G)$, then G is η'-choosable. In particular, if $\eta(v) \leq k - 1$ for each vertex v, $\eta(e) \leq k' - 1$ for each edge e', then G is (k, k')-choosable.*

Assume that A is a square matrix whose columns are expressed as linear combinations of columns of A_G. Define an index function $\eta_A : V(G) \cup E(G) \to \{0, 1, \ldots, \}$ as follows:

For $z \in V(G) \cup E(G)$, $\eta_A(z)$ is the number of columns of A in which $A_G(z)$ appears with non-zero coefficient.

The column vectors of A_G are not linearly independent. Thus a column of A may be written as a linear combination of columns of A_G in different ways. So the index function η_A is not uniquely determined by the matrix A itself; instead, it is determined by how its columns are expressed as linear combinations of the columns of A_G. For simplicity, we use the notation η_A. However, whenever the index function η_A is used, we refer to an explicit expression of its columns as linear combination of columns of A_G which is clear from the context. The following is a corollary of Lemma 5.3.2.

Corollary 5.3.4 *Assume that* $\sigma : V(G) \cup E(G) \to \{0, 1, \ldots\}$ *is a mapping which assigns to each vertex and each edge of G a non-negative integer. If there is a square matrix A with non-zero permanent whose columns are expressed as linear combinations of columns of A_G such that for each $z \in V(G) \cup E(G)$, $\eta_A(z) \le \sigma(z)$, then there exists a mapping* $\eta' : V(G) \cup E(G) \to \{0, 1, \ldots\}$ *such that* $\operatorname{per}(A_G(\eta')) \ne 0$ *and* $\eta'(z) \le \sigma(z)$ *for each* $z \in V(G) \cup E(G)$.

5.4 Permanent index

Definition 5.4.1 *The* permanent index *of a matrix A, denoted by* $\operatorname{pind}(A)$, *is the minimum integer k for which the following holds: There is a square matrix A' with $\operatorname{per}(A') \ne 0$ such that each column of A' is a column of A, and each column of A occurs at most k times as a column of A'. If such an integer k does not exist, then* $\operatorname{pind}(A) = \infty$.

By Corollary 5.3.3, if $\operatorname{pind}(A_G) \le k$, then G is $(k + 1, k + 1)$-choosable. Thus the following conjecture, proposed in [64], is stronger than the $(2, 2)$-choosability conjecture.

Conjecture 5.4.2 *For any graph G,* $\operatorname{pind}(A_G) = 1$.

The following conjecture, proposed in [10], is stronger than the $(1, 3)$-choosability conjecture.

Conjecture 5.4.3 *For any graph G with no isolated edges,* $\operatorname{pind}(B_G) \le 2$.

Lemma 5.4.4 *Let $S = [s_{i,j}]$ denote the square matrix of order $2k + 1$, whose first two columns are identical, with one coordinate*

0, k *coordinates* 1 *and* k *coordinates* -1. *The other entries of* S *are equal to* 1, *i.e.,* $s_{1,1} = s_{1,2} = 0$, $s_{i,1} = s_{i,2} = -1$ *for* $i = 2, 3, \ldots, k+1$, *and* $s_{i,j} = 1$ *for other* (i,j). *Then* $\operatorname{per}(S) = -(2k)!$.

Proof. For $i = 1, 2, \ldots, 2k+1$, let $S_k(i)$ denote the submatrix of order $2k$ of S obtained by deleting the i-th row and first column of S. Expanding along the first column, we have

$$\operatorname{per}(S) = \sum_{i=1}^{n} \{ s_{i,1} \times \operatorname{per}(S_k(i)) \}.$$

If $s_{i,1} = 1$, then $k-1$ coordinates of the first column of $S_k(i)$ is 1, and k coordinates of the first column of $S_k(i)$ is -1. Expanding along the first column, we have $\operatorname{per}(S_k(i)) = (k-1)(2k-1)! + (-1)k(2k-1)! = -(2k-1)!$. If $s_{i,1} = -1$ then the first column of $S_k(i)$ has k entries each equal to 1, and $k-1$ entries each equal to -1. Hence $\operatorname{per}(S_k(i)) = k(2k-1)! + (-1)(k-1)(2k-1)! = (2k-1)!$. Therefore $\operatorname{per}(S_k) = -(2k)!$. ∎

Theorem 5.4.5 ([10]) *Assume that* $G = (V, E)$ *is a graph with* $\operatorname{pind}(B_G) \le 2$. *Let* U *be a nonempty subset of* $V(G)$. *Let* F *be the graph obtained by adding two new vertices* u, v *to* G *and joining them to each vertex of* U. *Let* H *be the graph obtained from* F *by joining* u *and* v *by an edge. Then* $\operatorname{pind}(B_F) \le 2$ *and* $\operatorname{pind}(B_H) \le 2$.

Proof. Set $U = \{u_1, u_2, \ldots, u_k\}$. Let $e_i = uu_i$ and $e_i' = vu_i$, $1 \le i \le k$. Since $\operatorname{pind}(B_G) \le 2$, $\eta : E(G) \to \{0, 1, 2\}$ has the property that $\operatorname{per}(B_G(\eta)) \ne 0$.

First we prove that $\operatorname{pind}(B_F) \le 2$. Orient the new edges from u to U and v to U. Let Q be the matrix obtained from $B_F(\eta)$ by adding two copies of each of the columns $A_F(e_i) - A_F(e_i')$. Then

$$Q = \left[\begin{array}{c|c} R & B \\ \hline 0 & B_G(\eta) \end{array} \right],$$

where

$$R = \begin{bmatrix} -1 & -1 & \cdots & -1 \\ 1 & 1 & \cdots & 1 \\ -1 & -1 & \cdots & -1 \\ 1 & 1 & \cdots & 1 \\ \vdots & \vdots & \cdots & \vdots \\ 1 & 1 & \cdots & 1 \end{bmatrix}.$$

Therefore by Lemma 5.4.4, $\operatorname{per}(Q) = \operatorname{per}(R)\operatorname{per}(B_G(\eta)) =$

$(-1)^k (2k)! \mathrm{per}(B_G(\eta)) \neq 0$. As each column of Q is a linear combination of columns of B_F and each column of B_F occurs at most twice in such linear combinations, it follows from Lemma 5.3.2 that $\mathrm{pind}(B_F) \leq 2$.

The proof for the assertion on the graph H is similar. Now we have one more edge uv. Orient it from v to u. The matrix B_H is obtained from B_F by adding one row and one column indexed by $e^* = uv$. Similarly, let $\eta : E(G) \to \{0, 1, 2\}$ be a map with $\mathrm{per}(B_G(\eta)) \neq 0$. Let Q be obtained from $B_H(\eta)$ by adding two copies of the column $B_H(e^*)$, one copy of $B_H(e_1) - B_H(e_1')$ and for $i = 2, 3, \ldots, k$, two copies of the columns $B_H(e_i) - B_H(e_i')$. Then

$$Q = \left[\begin{array}{c|c} R' & B \\ \hline 0 & B_G(\eta) \end{array} \right],$$

where R' is the square matrix of order $2k + 1$ given by

$$R' = \begin{bmatrix} 0 & 0 & 2 & 2 & \cdots & 2 \\ -1 & -1 & -1 & -1 & \cdots & -1 \\ -1 & -1 & 1 & 1 & \cdots & 1 \\ -1 & -1 & -1 & -1 & \cdots & -1 \\ \vdots & \vdots & \vdots & \vdots & \vdots & \\ -1 & -1 & 1 & 1 & \cdots & 1 \end{bmatrix}.$$

Let Q' be obtained from Q by multiplying the first row by $\frac{1}{2}$ and for $i = 1, 2, \ldots, k$, the $(2i)$th row by -1. Then

$$Q' = \left[\begin{array}{c|c} S_k & B' \\ \hline 0 & B_G(\eta) \end{array} \right],$$

where S_k is the matrix defined in Lemma 5.4.4. Now $\mathrm{per}(Q') = \mathrm{per}(S_k) \mathrm{per}(B_G(\eta)) \neq 0$. This implies that $\mathrm{per}(Q) \neq 0$. As each column of Q is a linear combination of columns of B_H and each column of B_H occurs at most twice in such linear combinations, it follows from Lemma 5.3.2 that $\mathrm{pind}(B_H) \leq 2$. ∎

Corollary 5.4.6 *If $G \neq K_2$ is a clique or a complete bipartite graph, then $\mathrm{pind}(B_G) \leq 2$. Hence G is $(1, 3)$-choosable.*

Proof. The proof is by induction on the number of vertices of G. If $G \in \{K_3, K_4, K_{1,2}, K_{1,3}, K_{2,2}\}$, then direct computation shows that $\mathrm{pind}(B_G) \leq 2$. If $G = K_n$ for some $n \geq 5$, or $G = K_{n,m}$ for some $n + m \geq 5$, then G has two vertices u, v, such that either $uv \notin E(G)$ and $N_G(u) = N_G(v)$ or $uv \in E(G)$ and $N_G[u] = N_G[v]$, and $G - \{u, v\}$ is a complete graph or a complete bipartite graph. The conclusion follows from Theorem 5.4.5. ∎

5.5 Trees with an even number of edges

There are many graphs that are not $(1,2)$-choosable. The following lemma can be used to construct infinitely many non-$(1,2)$-choosable graphs.

Lemma 5.5.1 *If G_1, G_2 are non-$(1,2)$-choosable, and G is obtained from the disjoint union of G_1 and G_2 by adding one edge $e = uv$ connecting a vertex $u \in V(G_1)$ and a vertex $v \in V(G_2)$, then G is non-$(1,2)$-choosable.*

Proof. For $i = 1, 2$, as G_i is not $(1,2)$-choosable, there is a $(1,2)$-list assignment L_i for G_i for which G_i has no proper L_i-total weighting. Let L be the $(1,2)$-list assignment of G defined as follows:

$$L(z) = \begin{cases} \{0,1\}, & \text{if } z = e, \\ L_1(z), & \text{if } z \in V(G_1) \cup E(G_1), \\ L_2(z), & \text{if } z \in V(G_2) \cup E(G_2) \text{ and } z \neq v, \\ \{w-1\}, & \text{if } z = v \text{ and } L_2(v) = \{w\}. \end{cases}$$

We now show that G has no proper L-total weighting. Assume to the contrary that f is a proper L-total weighting of G. If $f(e) = 0$, then the restriction of f to G_1 is a proper L_1-total weighting of G_1, contrary to our assumption. If $f(e) = 1$, then let f' be the restriction of f to G_2, except that $f'(v) = f(v) + 1$. Then f' is a proper L_2-total weighting of G_2, again contrary to our assumption. ∎

We know that K_2 is non-$(1,2)$-choosable. Starting from K_2, by repeatedly using Lemma 5.5.1, we can construct infinitely many non-$(1,2)$-choosable trees. The question as to which trees are $(1,2)$-choosable was studied in [16]. An algorithm is presented in [16] which determines whether a given tree is $(1,2)$-choosable in linear time. The algorithm is a little complicated one.

All the non-$(1,2)$-choosable trees constructed by repeatedly using Lemma 5.5.1 have an odd number of edges. This is not a coincidence. In the following, we prove that the parity of $\operatorname{per}(B_T)$ for a tree T is different from the parity of the number of edges of T. As a consequence, if T is a tree with an even number of edges, then $\operatorname{per}(B_T) \neq 0$ and so T is $(1,2)$-choosable.

Lemma 5.5.2 *Assume that G has a cut edge e and G_1, G_2 are the two components of $G - e$. If G_i' is obtained from G_i by adding the edge e, then*

$$\mathrm{per}(B_G) = \mathrm{per}(B_{G_1})\,\mathrm{per}(B_{G_2'}) + \mathrm{per}(B_{G_1'})\,\mathrm{per}(B_{G_2}).$$

Proof. Assume that the edges of G_1 are $e_1, e_2, \ldots, e_{t-1}$, the cut edge $e = e_t$, and the edges of G_2 are $e_{t+1}, e_{t+2}, \ldots, e_m$. Then

$$
B_G = \begin{pmatrix}
 & & & * & & & \\
 & B_{G_1} & & * & & \mathbf{0} & \\
 & & & * & & & \\
* & * & * & 0 & * & * & * \\
 & & & * & & & \\
 & \mathbf{0} & & * & & B_{G_2} & \\
 & & & * & & &
\end{pmatrix}
$$

where the row and the column containing $*$'s are row t and column t, respectively, and a $*$ represents an unknown number in $\{-1, 0, 1\}$. Let u^1 be the row vector whose first $t - 1$ coordinates agree with row t of B_G and the last $m - t$ coordinates are 0. Let u^2 be the row vector whose first $t - 1$ coordinates are 0 and the last $m - t$ coordinates agree with row t of B_G. So row t of B_G is the sum $u^1 + u^2$. Let B_i be obtained from B_G by replacing row t with u^i, i.e.,

$$
B_1 = \begin{pmatrix}
 & & & * & & & \\
 & B_{G_1} & & * & & \mathbf{0} & \\
 & & & * & & & \\
0 & 0 & 0 & 0 & * & * & * \\
 & & & * & & & \\
 & \mathbf{0} & & * & & B_{G_2} & \\
 & & & * & & &
\end{pmatrix} \text{ and}
$$

$$
B_2 = \begin{pmatrix}
 & & & * & & & \\
 & B_{G_1} & & * & & \mathbf{0} & \\
 & & & * & & & \\
* & * & * & 0 & 0 & 0 & 0 \\
 & & & * & & & \\
 & \mathbf{0} & & * & & B_{G_2} & \\
 & & & * & & &
\end{pmatrix}.
$$

By linearity of permanent with respect to row vectors, we have

$$\mathrm{per}(B_G) = \mathrm{per}(B_1) + \mathrm{per}(B_2).$$

The non-zero contribution to per(B_1) from rows $t, t+1, \ldots, m$ are from columns $t, t+1, \ldots, m$. So per(B_1) = per(B_{G_1}) per($B_{G_2'}$). Similarly we have per(B_2) = per($B_{G_1'}$) per(B_{G_2}). Here $B_{G_1'}$ (resp. $B_{G_2'}$) stands for the matrix B_{G_1} (resp. B_{G_2}) augmented by the upper-left part of the t-th row and the t-th column of B_2 (resp. by the lower-right part of the t-th row and the t-th column of B_1). Hence

$$\mathrm{per}(B_G) = \mathrm{per}(B_{G_1})\,\mathrm{per}(B_{G_2'}) + \mathrm{per}(B_{G_1'})\,\mathrm{per}(B_{G_2}). \qquad \blacksquare$$

Lemma 5.5.3 *If T is a tree with m edges, then* per(B_T) $\equiv (m-1)$ (mod 2).

Proof. Since we are only concerned about the modulo 2 value of the permanent, the calculations can be done in Z_2. Hence we can replace -1 by 1 in B_T, and the resulting matrix is just the adjacency matrix of the line graph of G.

We prove the lemma by induction on m. If T is a star, then $B_T = J_m - I_m$, where J_m is the $m \times m$ matrix with all the entries equal to 1 and I_m is the $m \times m$ identity matrix. Hence per(B_T) is the number of permutations of $\{1, 2, \ldots, m\}$ with no fixed point, which is called *the mth derangement number* and denoted by d_m. It is easy to check that $d_1 = 0, d_2 = 1$ and $d_m = (m-1)(d_{m-1}+d_{m-2})$ for $m \geq 3$. Reducing modulo 2, by induction, we have $d_m \equiv (m-1)$ (mod 2).

Assume that T is not a star. Let e be any edge of T so that each of the two components T_1 and T_2 of $T - e$ has at least one edge. By Lemma 5.5.2,

$$\mathrm{per}(B_T) = \mathrm{per}(B_{T_1})\,\mathrm{per}(B_{T_2'}) + \mathrm{per}(B_{T_1'})\,\mathrm{per}(B_{T_2}),$$

where T_1, T_2, T_1', T_2' are as defined in Lemma 5.5.2. Assume that T_1 has $t-1$ edges and T_2 has $m-t$ edges. By applying the induction hypothesis to each of T_1, T_1', T_2, T_2', we conclude that

$$\mathrm{per}(B_T) \equiv (t-2)(m-t) + (t-1)(m-t-1) \equiv (m-1) \quad (\mathrm{mod}\ 2).$$

$$\blacksquare$$

Corollary 5.5.4 *If T is a tree with an even number of edges, then T is $(1,2)$-choosable.*

Trees with an odd number of edges may be non-$(1,2)$-choosable. Nevertheless, all trees with at least two edges are $(1,3)$-choosable.

Lemma 5.5.5 *If G' is obtained from G by adding two new vertices u, v and two new edges $e_1 = uv, e_2 = uw$, where $w \in V(G)$, then $\mathrm{pind}(B_{G'}) \leq \mathrm{pind}(B_G)$.*

Proof. Assume that $\mathrm{pind}(B_G) = k$ and $\eta : E(G) \to \{0, 1, 2, \ldots, k\}$ is a mapping such that $\mathrm{per}(B_G(\eta)) \neq 0$. Let $\eta' : E(G') \to \{0, 1, 2 \ldots, k\}$ be defined as $\eta'(e_1) = \eta'(e_2) = 1$ and $\eta'(e) = \eta(e)$ for $e \in E(G)$. As the row indexed by e_1 has only one non-zero entry in the column indexed by e_2, and the column indexed by e_1 has only one non-zero entry in the row indexed by e_2, we have $\mathrm{per}(B_{G'}(\eta')) = \mathrm{per}(B_G(\eta)) \neq 0$. Hence $\mathrm{pind}(B_{G'}) \leq k$. ∎

Theorem 5.5.6 *If T is a tree with at least two edges, then $\mathrm{pind}(B_T) \leq 2$, and hence T is $(1, 3)$-choosable.*

Proof. The proof is by induction on the number of vertices. If T has at most 4 edges, then this can be checked directly. Assume that T has at least 5 edges. Then either T has two leaves sharing a common neighbour or T has a pendant edge $e = uv$ (i.e., v is a leaf and u has degree 2 in T). In the former case, it follows from Theorem 5.4.5 and the induction hypothesis that $\mathrm{pind}(B_T) \leq 2$. In the latter case, it follows from Lemma 5.5.5 and the induction hypothesis that $\mathrm{pind}(B_T) \leq 2$. ∎

5.6 Complete graphs

In this section, we prove that the complete graphs K_n are $(2, 2)$-choosable.

Theorem 5.6.1 *For each $n \geq 1$, the complete graph K_n is $(2, 2)$-choosable.*

Proof. Let $V(K_n) = \{v_1, \ldots, v_n\}$. Denote by e_{ij} the edge $v_i v_j$. Let $\eta_n : V(K_n) \cup E(K_n) \to \{0, 1\}$ be defiend as

$$\eta_n(z) = \begin{cases} 0, & \text{if } z = v_1, e_{12}, e_{23}, \ldots, e_{(n-1)n}, \\ 1, & \text{otherwise.} \end{cases}$$

To prove Theorem 5.6.1, it suffices to show that $\mathrm{per}(A_{K_n}(\eta_n)) \neq 0$, which follows from Lemma 5.6.2.

Lemma 5.6.2 *For any positive integer* $n \geq 2$, $\mathrm{per}(A_{K_n}(\eta_n)) = \psi(n-1)$, *where* $\psi(1) = 1$ *and* $\psi(i) = \psi(i-1)i!$ *for* $i \geq 2$.

We prove this lemma by induction on $n \geq 2$. The case $n = 2$ is trivial. For $n = 3$,

$$A_{K_3} = \begin{bmatrix} 1 & -1 & 0 & 0 & 1 & -1 \\ 1 & 0 & -1 & 1 & 0 & -1 \\ 0 & 1 & -1 & 1 & -1 & 0 \end{bmatrix}$$

and

$$A_{K_3}(\eta_3) = \begin{bmatrix} -1 & 0 & 1 \\ 0 & -1 & 0 \\ 1 & -1 & -1 \end{bmatrix}.$$

It is easy to verify that $|\mathrm{per}(A_{K_3}(\eta_3))| = 2 = \psi(2)$. (The case $n = 3$ need not be considered separately. We did it explicitly for illustrating the lemma). It remains to show that if $n \geq 4$ and Lemma 4.5.2 holds for $n - 1$, it also holds for n.

Assume that I is a non-empty subset of $\{1, 2, \ldots, n-2\}$. Let $\eta_{n,I} : V(K_n) \cup E(K_n) \to \{0, 1, \ldots, \}$ be defined as follows:

$$\eta_{n,I}(z) = \begin{cases} 0, & \text{if } z = e_{12}, e_{23}, \ldots, e_{(n-1)n}, e_{(n-2)n}, \\ & \quad \ldots, e_{1n} \text{ or } z = v_1 \text{ and } 1 \notin I, \\ 2, & \text{if } z = v_i \text{ and } i \in I - \{1\}, \\ n - 1 - |I|, & \text{if } z = v_n, \\ 1, & \text{otherwise.} \end{cases}$$

Straightforward calculation shows that $\sum_{z \in V(K_n) \cup E(K_n)} \eta_{n,I}(z) = |E(K_n)| = n(n-1)/2$. So $A_{K_n}(\eta_{n,I})$ is a square matrix.

Lemma 5.6.3 *For any non-empty subset* I *of* $\{1, 2, \ldots, n-2\}$, $\mathrm{per}(A_{K_n}(\eta_{n,I})) = 0$.

Proof. Assume to the contrary that $\mathrm{per}(A_{K_n}(\eta_{n,I})) \neq 0$. By Corollary 5.3.3, K_n has an L-total weighting for any total list assignment L for which $|L(z)| = \eta(z) + 1$ for each $z \in V \cup E$. Let L be the total list assignment defined as $L(z) = \{0, 1, \ldots, \eta_{n,I}(z)\}$ for each $z \in V \cup E$.

Let f be an L-total weighting of K_n. Consider the induced colouring ϕ_f of K_n defined as $\phi_f(v) = f(v) + \sum_{e \in E(v)} f(e)$. If $v = v_1$, or $v = v_{n-1}$, then v is incident to two edges whose weights are 0: For v_1, the two edges are e_{12}, e_{1n}. For v_{n-1}, the two edges are $e_{(n-2)(n-1)}, e_{(n-1)n}$. Every other edge incident to v has weight 0 or 1, and vertex v itself has weight 0 or 1. Therefore $\phi_f(v) \in \{0, 1, \ldots, n-2\}$. If $v = v_j$ for some $2 \leq j \leq n-2$, then v is incident to three edges $e_{j(j-1)}, e_{(j+1)j}, e_{nj}$, of weight 0. The other edges have weights 0 or 1, and v_j itself has weight 0, 1 or 2. So again $\phi_f(v) \in \{0, 1, \ldots, n-2\}$. All the edges incident to v_n have weight 0, and v_n itself has weight in $\{0, 1, \ldots, n-1-|I|\}$. As $|I| \geq 1$, so $\phi_f(v_n) \in \{0, 1, \ldots, n-2\}$. Therefore ϕ_f is not a proper colouring of K_n, contrary to the assumption that f is an L-total weighting of K_n. ∎

Now we are ready to prove Lemma 5.6.2. Let $m = n(n-1)/2$ and $m' = (n-1)(n-2)/2 = m - n + 1$. Then $A_{K_n}(\eta_n)$ is an $m \times m$ matrix. The submatrix of $A_{K_n}(\eta_n)$ induced by the vertices and edges in K_{n-1} is $A_{K_{n-1}}(\eta_{n-1})$. The other $n-1$ columns are $A_{K_n}(e_{1n}), A_{K_n}(e_{2n}), \ldots, A_{K_n}(e_{(n-2)n}), A_{K_n}(v_n)$. By Lemma 5.1.2, for $j = 1, 2, \ldots, n-2$, $A_{K_n}(e_{jn}) = A_{K_n}(v_n) + A_{K_n}(v_j)$.

For each subset I of $\{1, 2, \ldots, n-2\}$, let Q_I be the matrix obtained from $A_{K_n}(\eta_n)$ by the following operations:

- For $j \in I$, replace the column $A_{K_n}(e_{jn})$ with column $A_{K_n}(v_j)$.

- For $j \in \{1, 2, \ldots, n-2\} - I$, replace the column $A_{K_n}(e_{jn})$ with column $A_{K_n}(v_n)$.

By Lemma 5.1.2,

$$\mathrm{per}(A_{K_n}(\eta_n)) = \sum_{I \subseteq \{1,2,\ldots,n-2\}} \mathrm{per}(Q_I).$$

Note that if $I = \emptyset$, then Q_I has $n-1$ columns indexed by v_n, i.e., Q_I contains $n-1$ copies of the column $A_{K_n}(v_n)$, and $n-1$ rows indexed by edges incident to v_n. By deleting these $n-1$ rows and the $n-1$ columns, the resulting matrix is $A_{K_{n-1}}(\eta_{n-1})$. The submatrix of Q_I with rows indexed by the $n-1$ edges incident to v_n and with columns indexed by the $n-1$ copies of v_n is an $(n-1) \times (n-1)$ matrix all of whose entries are 1. Therefore

$$\mathrm{per}(Q_I) = \mathrm{per}(A_{K_{n-1}})(n-1)! = \psi(n-2)(n-1)! = \psi(n-1).$$

For each non-empty subset I of $\{1, 2, \ldots, n-2\}$, $Q_I = A_{K_n}(\eta_{n,I})$. By Lemma 5.6.3, $\mathrm{per}(Q_I) = 0$. Therefore, $\mathrm{per}(A_{K_n}(\eta_n)) = \psi(n-1)$. This completes the proof of Lemma 5.6.2. ∎

5.7 Every graph is (2,3)-choosable

In this section, we prove that every graph is $(2,3)$-choosable [65]. By Corollary 5.3.3, it suffices to prove the following theorem.

Theorem 5.7.1 *For every graph G, there exists a mapping η : $V(G) \cup E(G) \to \{0, 1, \ldots\}$ with $\eta(v) \leq 1$ for $v \in V(G)$, $\eta(e) \leq 2$ for $e \in E(G)$ and $\operatorname{per}(A_G(\eta)) \neq 0$.*

We shall prove a slightly stronger result.

Theorem 5.7.2 *Assume that G is a connected graph and that F is a spanning tree of G. Then there is a matrix A whose columns are linear combinations of columns of A_G such that $\operatorname{per}(A) \neq 0$ and $\eta_A(v) \leq 1$ for $v \in V$, $\eta_A(e) = 0$ for $e \in E(F)$ and $\eta_A(e) \leq 2$ for $e \in E(G) - E(F)$.*

Proof. The proof is by contradiction. Assume that the result is not true, and that G is a minimum counterexample. It is obvious that $|V| \geq 3$.

Let u be a vertex of G, which is a leaf of F. Assume that $N_G(u) = \{u_1, u_2, \ldots, u_k\}$. Assume that the edge $u_k u \in E(F)$. Let $G' = G - u$. By the minimality of G, G' has an index function η' such that $\operatorname{per}(A_{G'}(\eta')) \neq 0$, and $\eta'(v) \leq 1$ for $v \in V(G')$, $\eta'(e) \leq 2$ for $e \in E(G')$, and moreover, for $e \in E(F - u)$, $\eta'(e) = 0$.

Let $|E(G')| = m'$ where $m' = |E(G)| - k = m - k$. We view η' as an index function of G, with $\eta'(z) = 0$ if $z \in (V(G) \cup E(G)) - (V(G') \cup E(G'))$. Then $A_G(\eta')$ is an $m \times m'$ matrix, consisting of m' columns of A_G. Let $\eta = \eta'$ except that $\eta(u) = k$. Then $A_G(\eta)$ is an $m \times m$ matrix, which is obtained from $A_G(\eta')$ by adding k copies of the column $A_G(u)$. The added k columns have k rows (the rows indexed by edges incident to u) that are all 1's, and all the other entries of these k columns are 0. Therefore $\operatorname{per}(A_G(\eta)) = \operatorname{per}(A_{G'}(\eta'))k!$, and hence $\operatorname{per}(A_G(\eta)) \neq 0$.

Let $M_0 = A_G(\eta)$. For $i = 1, 2, \ldots, k-1$, if $\eta'(u_i) = 0$, then let $M_i = M_{i-1}$. If $\eta'(u_i) = 1$, then let M_i be obtained from M_{i-1} by replacing $A_G(u_i)$ with $A_G(e_i)$.

Claim 5.7.3 *For $i = 1, 2, \ldots, k-1$, $\operatorname{per}(M_i) = \operatorname{per}(M_{i-1})$.*

Proof. If $\eta'(u_i) = 0$, then $M_i = M_{i-1}$, and there is nothing to prove. Assume that $\eta'(u_i) = 1$ and M_i is obtained from M_{i-1} by replacing $A_G(u_i)$ with $A_G(e_i)$. Let M_i' be obtained from M_{i-1} by replacing $A_G(u_i)$ with $A_G(u)$. In M_i', the column $A_G(u)$ occurs $k+1$ times. These $k+1$ columns have k rows (the rows indexed by edges incident to u) that are all 1's, and all the other entries of these $k+1$ columns are 0. Therefore per$(M_i') = 0$. By Lemma 5.3.1, $A_G(e_i) = A_G(u_i) + A_G(u)$. By Lemma 5.1.2, per$(M_i) = $ per$(M_{i-1}) + $ per$(M_i') = $ per(M_{i-1}). ∎

Observe that $M_{k-1} = A_G(\eta)$ for an index function η of G for which the following hold:

- $\eta(u_i) = 0$ for $i = 1, 2, \ldots, k-1$, $\eta(u) = k$, $\eta(v) \leq 1$ for other vertices v of G.

- $\eta(e_i) \leq 1$ for $i = 1, 2, \ldots, k-1$, $\eta(e) = 0$ for edges in F, and $\eta(e) \leq 2$ for other edges of G.

By Lemma 5.3.1, $A_G(u) = A_G(e_i) - A_G(u_i)$ for each $i \in \{1, 2, \ldots, k-1\}$. We replace $k-1$ copies of $A_G(u)$ with $A_G(e_i) - A_G(u_i)$ $(i = 1, 2, \ldots, k-1)$. Denote the resulting matrix by A. The matrix A is indeed the same as $A_G(\eta)$. However, in this new format, we have $\eta_A(v) \leq 1$ for all vertices v of G, and $\eta_A(e) \leq 2$ for all edges of G, and $\eta_A(e) = 0$ for $e \in E(F)$. As per$(A) = $ per$(A_G(\eta)) \neq 0$, this completes the proof of Theorem 5.7.2. ∎

Theorem 5.7.4 *Every graph is $(2,3)$-choosable.*

A result slightly stronger than Theorem 5.7.4 follows from Theorem 5.7.2: Suppose that G is a connected graph and F is a spanning tree of G. Let $\psi : V(G) \cup E(G) \rightarrow \{1, 2, 3\}$ be defined as $\psi(v) = 2$ for every vertex v, $\psi(e) = 1$ for $e \in E(F)$ and $\psi(e) = 3$ for $e \in E(G) - E(F)$. Then G is total weight ψ-choosable. The non-list version of this result was obtained by Kalkowski [31].

As a weaker version of the (1,3)-choosability conjecture, it was conjectured in [64] as there is a constant k such that every graph with no isolated edges is (1,k)-choosable. This conjecture was confirmed in [15], where it was proved that every graph with no isolated edges is (1,17)-choosable, and this result has been improved in [69] that every graph with no isolated edges is (1,5)-choosable. The following conjecture is weaker than the (1,2)-choosability conjecture and remains open.

Conjecture 5.7.5 *There is a constant k such that every graph is $(k,2)$-choosable.*

Bibliography

[1] N. Alon, S. Friedland, and G. Kalai. Regular subgraphs of almost regular graphs. *J. Combin. Theory Ser. B*, 37(1):79–91, 1984.

[2] N. Alon and M. Tarsi. A nowhere-zero point in linear mappings. *Combinatorica*, 9(4):393–395, 1989.

[3] N. Alon and M. Tarsi. Colourings and orientations of graphs. *Combinatorica*, 12(2):125–134, 1992.

[4] N. Alon. Combinatorial Nullstellensatz. *Combin. Probab. Comput.*, 8(1-2):7–29, 1999. Recent trends in combinatorics (Mátraháza, 1995).

[5] N. Alon and Z. Füredi. Covering the cube by affine hyperplanes. *European J. Combin.*, 14(2):79–83, 1993.

[6] M. Artin. *Algebra*. Prentice Hall, Inc., Englewood Cliffs, NJ, 1991.

[7] R. Balakrishnan and K. Ranganathan. *A Textbook of Graph Theory*. Springer, Second Edition, 2012.

[8] R. Balakrishnan and Sriraman Sridharan. *Discrete Mathematics: Graph Algorithms, Algebraic Structures, Coding Theory and Cryptography*. CRC Press (Taylor and Francis), 2019.

[9] É. Balandraud and B. Girard. A Nullstellensatz for sequences over $\mathbb{F}_p$. *Combinatorica*, 34(6):657–688, 2014.

[10] T. Bartnicki, J. Grytczuk, and S. Niwczyk. Weight choosability of graphs. *J. Graph Theory*, 60(3):242–256, 2009.

[11] G. Batzaya and G. Bayarmagnai. A generalized Combinatorial Nullstellensatz for multisets. *European J. Combin.*, 83:103013, 2020.

[12] A. Bishnoi, P. L. Clark, A. Potukuchi, and J. R. Schmitt. On zeros of a polynomial in a finite grid. *Combin. Probab. Comput.*, 27(3):310–333, 2018.

[13] B. Bosek, J. Grytzcuk, G. Gutowski, and O. Serra. A note on counting solutions in algebraic method. Private communication, 2019.

[14] L. Cai, W. Wang and X. Zhu. Choosability of toroidal graphs without short cycles. *J. Graph Theory*, 65(1):1–15, 2010.

[15] L. Cao, *Total weight choosability of graphs: Towards the 1-2-3 conjecture*, J. Combin. Theory Ser. B 149 (2021), 109-146.

[16] G. Chang, G. Duh, T. Wong, and X. Zhu. Total weight choosability of trees. *SIAM J. Discrete Math.*, 31(2):669–686, 2017.

[17] H. Choi and Y. Kwon. On t-common list-colourings. *Electron. J. Combin.*, 24(3):Paper 3.32, 10, 2017.

[18] L. J. Cowen, R. H. Cowen, and D. R. Woodall. Defective colourings of graphs in surfaces: partitions into subgraphs of bounded valency. *J. Graph Theory*, 10(2):187–195, 1986.

[19] W. Cushing and H. A. Kierstead. Planar graphs are 1-relaxed, 4-choosable. *European Journal of Combinatorics*, 31(5):1385–1397, 2010.

[20] L. Duraj, G. Gutowski, and J. Kozik. Chip games and paintability. *Electron. J. Combin.*, 23(3):Paper 3.3, 12, 2016.

[21] N. Eaton and T. Hull. Defective list colourings of planar graphs. *Bulletin of the Institute of Combinatorics and its Applications*, 25:79–87, 1999.

[22] M. N. Ellingham and Luis Goddyn. List edge colourings of some 1-factorable multigraphs. *Combinatorica*, 16(3):343–352, 1996.

[23] P. Erdős, A. L. Rubin, and H. Taylor. Choosability in graphs. In *Proceedings of the West Coast Conference on Combinatorics, Graph Theory and Computing (Humboldt State Univ., Arcata, Calif., 1979)*, Congress. Numer., XXVI, pages 125–157. Utilitas Math., Winnipeg, Man., 1980.

[24] H. Fleischner and M. Stiebitz. Some remarks on the cycle plus triangles problem. In *The mathematics of Paul Erdős, II*, volume 14 of *Algorithms Combin.*, pages 136–142. Springer, Berlin, 1997.

[25] H. Fleischner and M. Stiebitz. A solution to a colouring problem of P. Erdős. *Discrete Math.*, 101(1-3):39–48, 1992. Special volume to mark the centennial of Julius Petersen's "*Die Theorie der regulären Graphs*", Part II.

[26] J. Grytczuk and X. Zhu. Alon–Tarsi number of planar graphs minus a matching. *J. Combin. Theory Ser. B*, 145: 511–520, 2020.

[27] G. Gutowski, M. Han, T. Krawczyk, and X. Zhu. Defective 3-paintability of planar graphs. *Electron. J. Combin.*, 25(2):Paper 2.34, 20, 2018.

[28] J. Hladký, D. Král, and U. Schauz. Brooks' theorem via the Alon–Tarsi theorem. *Discrete Math.*, 310(23):3426–3428, 2010.

[29] P. Huang, T. Wong, and X. Zhu. Application of polynomial method to on-line list colouring of graphs. *European J. Combin.*, 33(5):872–883, 2012.

[30] F. Jaeger. Problem presented in the 6th Hungar. Comb. Coll. *Finite and Infinite Sets (eds.: Hajnal, A., Lovász, L., Sös, V. T.).*, 1982.

[31] M. Kalkowski. personal communication via Grytczuk. 2009.

[32] M. Kalkowski, M. Karoński, and F. Pfender. Vertex-colouring edge-weightings: towards the 1-2-3-conjecture. *J. Combin. Theory Ser. B*, 100(3):347–349, 2010.

[33] M. Karonski, T. Łuczak, and A. Thomason. Edge weights and vertex colours. *J. Combin. Theory Ser. B*, 91(1):151–157, 2004.

[34] H. Kaul and J. A. Mudrock. On the Alon–Tarsi number and chromatic-choosability of Cartesian products of graphs. *Electron. J. Combin.*, 26(1):Paper 1.3, 13, 2019.

[35] R. Kim, S. Kim, and X. Zhu. The Alon–Tarsi number of subgraphs of a planar graph. 2019.

130   *Bibliography*

[36] S.Kim, Y. Kwon, and B. Park. Chromatic-choosability of the power of graphs. *Discrete Appl. Math.*, 180:120–125, 2015.

[37] S. Kim and B. Park. Bipartite graphs whose squares are not chromatic-choosable. *Electron. J. Combin.*, 22(1):Paper 1.46, 12, 2015.

[38] S. Kim and B. Park. Counterexamples to the list square colouring conjecture. *J. Graph Theory*, 78(4):239–247, 2015.

[39] G. Kós, T. Mészáros, and L. Rónyai. Some extensions of Alon's Nullstellensatz. *Publ. Math. Debrecen*, 79(3-4):507–519, 2011.

[40] G. Kós and L. Rónyai. Alon's Nullstellensatz for multisets. *Combinatorica*, 32(5):589–605, 2012.

[41] N. Kosar, S. Petrickova, B. Reiniger, and E. Yeager. A note on list-colouring powers of graphs. *Discrete Math.*, 332:10–14, 2014.

[42] A. V. Kostochka and D. R. Woodall. Choosability conjectures and multicircuits. *Discrete Math.*, 240(1-3):123–143, 2001.

[43] S. Lang. *Algebra*, volume 211 of *Graduate Texts in Mathematics*. Springer-Verlag, New York, third edition, 2002.

[44] M. Lasoń. A generalization of Combinatorial Nullstellensatz. *Electron. J. Combin.*, 17(1):Note 32, 6, 2010.

[45] Z. Li, Z. Shao, F. Petrov and A. Gordeev. The Alon–Tarsi number of toroidal grids, *arxiv: 1912.12466v1*.

[46] H. Lu and X. Zhu. The Alon–Tarsi number of planar graphs without cycles of lengths 4 and *l*. Discrete Mathematics, to appear.

[47] M. Michałek. A short proof of Combinatorial Nullstellensatz. *Amer. Math. Monthly*, 117(9):821–823, 2010.

[48] M. Mirzakhani. A small non-4-choosable planar graph. *Bull. Inst. Combin. Appl.* 17 (1996), 15–18.

[49] F. Petrov. General parity result and cycle-plus-triangles graphs. *J. Graph Theory*, 85(4):803–807, 2017.

[50] J. Przybyło and M. Woźniak. On a 1, 2 conjecture. *Discrete Math. Theor. Comput. Sci.*, 12(1):101–108, 2010.

[51] J. Przybyło and M. Woźniak. Total weight choosability of graphs. *Electron. J. Combin.*, 18(1):Paper 112, 11, 2011.

[52] R. Ramamurthi and D. B. West. Hypergraph extension of the Alon–Tarsi list colouring theorem. *Combinatorica*, 25(3):355–366, 2005.

[53] U. Schauz. Mr. Paint and Mrs. Correct. *Electronic Journal of Combinatorics*, 16(1):R77:1–18, 2009.

[54] U. Schauz and T. Honold. The Combinatorial Nullstellensatz in Balandraud and Girard's Theorem and in Alon Tarsi's Theorem. *Manuscript*, 2019.

[55] U. Schauz. Algebraically solvable problems: describing polynomials as equivalent to explicit solutions. *Electron. J. Combin.*, 15(1):Research Paper 10, 35, 2008.

[56] U. Schauz. Proof of the list edge colouring conjecture for complete graphs of prime degree. *Electron. J. Combin.*, 21(3):Paper 3.43, 17, 2014.

[57] D. E. Scheim. The number of edge 3-colourings of a planar cubic graph as a permanent. *Discrete Math.*, 8:377–382, 1974.

[58] R. Škrekovski. List improper colourings of planar graphs. *Combinatorics, Probability and Computing*, 8(3):293–299, 1999.

[59] V. A. Tashkinov. 3-regular subgraphs of 4-regular graphs. *Mat. Zametki*, 36(2):239–259, 1984.

[60] C. Thomassen. Every planar graph is 5-choosable. *Journal of Combinatorial Theory, Series B*, 62(1):180–181, 1994.

[61] V. G. Vizing. colouring the vertices of a graph in prescribed colours. *Diskret. Analiz*, (29 Metody Diskret. Anal. v Teorii Kodov i Shem):3–10, 101, 1976.

[62] M. Voigt and B. Wirth. On 3-colourable non-4-choosable planar graphs. *J. Graph Theory*, 24(3):233–235, 1997.

[63] D. B. West. *Introduction to graph theory*. Prentice Hall, Inc., Upper Saddle River, NJ, 1996.

[64] T. Wong and X. Zhu. Total weight choosability of graphs. *J. Graph Theory*, 66(3):198–212, 2011.

[65] T. Wong and X. Zhu. Every graph is $(2,3)$-choosable. *Combinatorica*, 36(1):121–127, 2016.

[66] L. Zhang. 4-regular graphs without 3-regular subgraphs. In *Graph theory and its applications: East and West (Jinan, 1986)*, volume 576 of *Ann. New York Acad. Sci.*, pages 691–699. New York Acad. Sci., New York, 1989.

[67] X. Zhu. Alon–Tarsi number of planar graphs. *Journal of Combinatorial Theory Ser. B*, https://doi.org/10.1016/j.jctb.2018.06.004.

[68] X. Zhu. Circular chromatic number: a survey. *Discrete Math.*, 229(1-3):371–410, 2001. Combinatorics, graph theory, algorithms and applications.

[69] X. Zhu, *Every nice graph is (1,5)-choosable*, arxiv.org/abs/2104.05410.

Index